3D Modeling of Nonlinear Wave Phenomena on Shallow Water Surfaces

Scrivener Publishing
100 Cummings Center, Suite 541J
Beverly, MA 01915-6106

Publishers at Scrivener
Martin Scrivener (martin@scrivenerpublishing.com)
Phillip Carmical (pcarmical@scrivenerpublishing.com)

3D Modeling of Nonlinear Wave Phenomena on Shallow Water Surfaces

Iftikhar B. Abbasov

Scrivener Publishing

WILEY

This edition first published 2018 by John Wiley & Sons, Inc., 111 River Street, Hoboken, NJ 07030, USA and Scrivener Publishing LLC, 100 Cummings Center, Suite 541J, Beverly, MA 01915, USA

For more information about Scrivener publications please visit www.scrivenerpublishing.com.

Wiley Global Headquarters
111 River Street, Hoboken, NJ 07030, USA

For details of our global editorial offices, customer services, and more information about Wiley products visit us at www.wiley.com.

Library of Congress Cataloging-in-Publication Data

ISBN 978-1-119-48796-8

Cover image: Kris Hackerott
Cover design by Kris Hackerott

Set in size of 11pt and Minion Pro by Exeter Premedia Services Private Ltd., Chennai, India

10 9 8 7 6 5 4 3 2 1

Contents

Preface

How mesmerizing is the beauty of the waves approaching the seashore against a background of the sunset: they try to catch up with each other in a continuous cycle of water flow, then they subside, then intensify, rolling up on the shore, crashing into a sparkling foam, creating an endless symphony of surf. You can endlessly admire this landscape, which has existed for billions of years, from the time when there were no living beings on the planet Earth. Also, primeval ocean waves wash ashore, as is happening now in the presence of a person watching this picture. These waves have attracted the attention of artists and researchers for more than a century. Despite their beauty and simplicity, however, they are not always easy to describe. Moreover, to verify the plausibility of the created model, special knowledge is not necessarily required. It's enough to go to the beach, and everything will become clear.

At the same time, neglecting the power of this beauty can lead to devastating consequences in storm surges and earthquakes. Therefore, the study of waves on the sea surface is not an easy task, and attempts are made in this work to describe and simulate some wave events on the surface of the aquatic environment. By their nature, these waves are inherently nonlinear, although some approximations may be considered linear. Consequently, the most appropriate theory of surface wave description is nonlinear theory.

This book presents the work done by the author for the research and modeling of nonlinear wave activities on the shallow water surface. An attempt was made to describe the run-up of surface waves to various coastal formations in shallow waters. Photographic illustrations of wave activities on the shallow water surface, made by the author, are also provided to illustrate the work.

I want to express my appreciation to my teachers, and promote a love for mathematics, art, and beauty.

Iftikhar B. Abbasov

Introduction

In the context of the study of the ecosystems of the shallow coastal areas of the world's oceans, physical phenomena occurring on the surface of the aquatic environment play an important role. These phenomena, like all natural phenomena, are complex and nonlinear. Therefore, this leads to the nonlinear mathematical models of the actual processes.

The theory of wave motion fluids is a classical section of hydrodynamics and has a three-hundred-year history. The interest in wave activities on the surface of the fluid could be explained by the prevalence and accessibility of this physical phenomenon. Despite a great deal of research, the theory of wave fluid movements is still incomplete.

Of great importance is the matter of researching and modeling the wave activities at shallow water and the impact of surface gravity waves to coast formations and hydrotechnical structures. Therefore, the question of 3D modeling of the distribution, run-up and refraction of nonlinear surface waves can play an important role in monitoring and forecasting the sustainable development of the ecosystems of these areas.

The results of the research and numerical modeling of the dynamic of nonlinear surface gravity waves at shallow water are introduced in this work. Corresponding equations of mathematical physics and methods of mathematical modeling are used for describing and modeling.

Analytical descriptions of these nonlinear wave activities often use different modifications of the shallow water equations. For the numerical modeling, shallow water equations are also used in a 1D case. 2D and 3D numerical modeling of nonlinear surface gravity waves to beach approaches are based on Navier-Stokes equations. Navier-Stokes equations allow for both nonlinear effects and turbulent processes to be considered in the incompressible fluid.

Therefore, appropriate nonlinear waves of hydrodynamic equations will be used to adequately model nonlinear wave activities in shallow water conditions.

1

Equations of Hydrodynamics

1.1 Features of the Problems in the Formulation of Mathematical Physics

When examining a physical process, the scientist needs to describe it in mathematical terms. A mathematical description or a process modeling could be quite varied. Mathematical modeling does not investigate the actual physical process itself, and some of its models are the ideal process written in the form of mathematics. The mathematical model should preserve the basic features of the actual physical process and, at the same time, should be simple enough to be solved by known methods. In the future, the consistency of the mathematical model with the actual process needs to be tested.

Many ways of mathematically describing physical processes lead to differential equations with private derivatives, and in some cases to Integro-differential equations. It is this group of tasks that is assigned the term mathematical physics, and the methods of solving them are referred to as mathematical physics methods.

The subject of mathematical physics is the mathematical theory of physical phenomena. The wide distribution of mathematical physics is connected to the commonality of mathematical models based on fundamental laws of nature: the laws of mass, energy, charge conservation, kinetic momentum. This results in the same mathematical models describing the physical phenomena of different natures.

Mathematical physics usually examines processes in a certain spatial area filled with a continuous material environment called the solid environment. Values that describe the state of the environment and the physical processes that occur in it depend on the spatial coordinates and time. In general, mathematical physics models describe the behavior of the system at three levels: the interaction of the system as a whole with the external environment; the interaction between the system's basic volumes and the properties of a single, basic system volume.

The interaction of the system with the external environment is the wording of the boundary conditions, i.e., the conditions at the border of the task area, which include in general the boundary and initial conditions. The second level describes the interaction of elementary volumes based on laws for the preservation of physical substances and their transfer in space. The third level corresponds to the establishment of the state equations of the environment, i.e., the creation of a mathematical model of the basic environment behavior.

The equations of mathematical physics emerged from the consideration of such essential physical tasks as the distribution of sound in gases, waves in liquids, heat in physical bodies. The phenomena of nuclear reaction, gravity, electromagnetic effects, the origin and evolution of the universe are being actively explored now. Mathematical models of these different physical phenomena lead to equations with private derivatives.

An equation with a private derivative is an equation that includes an unknown function that depends on several variables and its private derivatives. Dependence on many variables in an unknown function makes it much harder to solve equations with private derivatives. Very few of these equations are explicitly solved.

As a result of the development of computer technology, the role of computational methods in the approximation of mathematical physics has grown. However, the approximate analytical methods that make it possible to obtain the connection between the functions sought and the specified parameters of the task in question have not lost their importance.

A precise analytical solution to mathematical physics usually requires the integration of differential equations with private derivatives. These equations need to be integrated into a certain spatial-temporal area where the desired functions are subjected to the specified boundary conditions. Therefore, a precise analytical solution to such equations is possible only in rare cases, which underscores the importance of approximation methods. Before we go into the methods of solving equations, consider classifying differential equations with private derivatives.

1.2 Classification of Linear Differential Equations with Partial Derivatives of the Second Order

Many problems of mathematical physics lead to linear differential equations of the second order. For an unknown function u, a linear differential equation of the second order, depending on two variables x and y, has the following form [Aramanovich, 1969]:

$$A\frac{\partial^2 u}{\partial x^2}+B\frac{\partial^2 u}{\partial x\partial y}+C\frac{\partial^2 u}{\partial y^2}+D\frac{\partial u}{\partial x}+E\frac{\partial u}{\partial y}+Fu=f(x,y) \qquad (1.2.1)$$

We assume that all the coefficients of the equation are constant. Most differential equations of mathematical physics represent particular cases of the common equation (1.2.1).

L. Euler proved that any differential equation of the form (1.2.1) by replacing the independent variables x and y can be reduced to one of the following three types:

1. If, $AC-\frac{B^2}{4}>0$, then, after introducing new independent variables ξ and η equation (1.2.1) takes the form

   $$\frac{\partial^2 u}{\partial \xi^2}+\frac{\partial^2 u}{\partial \eta^2}+D_1\frac{\partial u}{\partial \xi}+E_1\frac{\partial u}{\partial \eta}+F_1 u=f_1(\xi,\eta) \qquad (1.2.2)$$

   In this case, the equation is called *elliptic*. The simplest elliptic equation is the Laplace equation.

2. If, $AC-\frac{B^2}{4}<0$, then equation (1.2.1) can be given the form

   $$\frac{\partial^2 u}{\partial \xi^2}-\frac{\partial^2 u}{\partial \eta^2}+D_2\frac{\partial u}{\partial \xi}+E_2\frac{\partial u}{\partial \eta}+F_2 u=f_2(\xi,\eta) \qquad (1.2.3)$$

Such an equation is called *hyperbolic;* the simplest example of this is the one-dimensional equation of free oscillations.

3. If, $AC - \frac{B^2}{4} = 0$, then equation (1.2.1) is reduced to the next:

$$\frac{\partial^2 u}{\partial \xi^2} + D_3 \frac{\partial u}{\partial \xi} + E_3 \frac{\partial u}{\partial \eta} + F_3 u = f_3(\xi, \eta) \quad (1.2.4)$$

This equation is called *parabolic.* An example of it is the equation of linear thermal conductivity.

The names of the equations are explained by the fact that in the study of the common equation of curves of the second order

$Ax^2 + Bxy + Cy^2 + Dx + Ey + F = 0$, it turns out that the curve represents:

in the case $AC - \frac{B^2}{4} > 0$ – *of an ellipse;*

in the case $AC - \frac{B^2}{4} < 0$ – *of an hyperbole;*

in the case $AC - \frac{B^2}{4} = 0$ – *of an parabola.*

Finally, any equation of the form (1.2.1) can be reduced to one of the following canonical types:

$$\frac{\partial^2 u}{\partial \xi^2} + \frac{\partial^2 u}{\partial \eta^2} + cu = f \quad (\textit{elliptical type}),$$

$$\frac{\partial^2 u}{\partial \xi^2} - \frac{\partial^2 u}{\partial \eta^2} + cu = f \quad (\textit{hyperbolic type}),$$

$$\frac{\partial^2 u}{\partial \xi^2} - \frac{\partial u}{\partial \eta} = f \quad (\textit{parabolic type}),$$

(c – constant number, f – function of variables x и y).

Equations of hyperbolic and parabolic types arise most often when studying processes occurring in time (equations of oscillations, wave propagation, heat propagation, diffusion). In the one-dimensional case, one coordinate always participates x and time t. Additional conditions for such tasks, divided into *initial and boundary.*

The initial conditions consist in setting for t=0 the values of the desired function u and its derivative (in the hyperbolic case) or only the values of the function itself (in the parabolic case).

The boundary conditions for these problems lie in the fact that the values of the unknown function $u(x,t)$ are indicated at the ends of the coordinate change interval.

If the process proceeds in an infinite interval of variation of the coordinate x, then the boundary conditions disappear, and the problem is obtained only with initial conditions, or, as it is often called, the *Cauchy problem.*

If a problem is posed for a finite interval, then the initial and boundary conditions must be given. Then we speak of a mixed problem.

Equations of elliptic type arise usually in the study of stationary processes. The time t does not enter into these equations, and both independent variables are the coordinates of the point. Such are the equations of the stationary temperature field, the electrostatic field, and the equations of many other physical problems. For problems of this type, only boundary conditions are set, that is, specifies the behavior of the unknown function on the contour area. This can be the Dirichlet problem, when the values of the function itself are given; the Neumann problem when the values of the normal derivative of the unknown function are given; and the problem, when a linear combination of the function is given on the contour, and its normal derivative.

In the basic problems of mathematical physics, it is physical considerations that prompt what additional conditions should be put in one or another problem in order to obtain a unique solution of it that corresponds to the nature of the process being studied.

In addition, it should be borne in mind that all the equations derived are of an idealized nature, that is, they reflect only the most essential features of the process. The functions entering into the initial and boundary conditions in physical problems are determined from experimental data and can be considered only approximately.

1.3 Nonlinear Equations of Fluid Dynamics

Linear integro-differential equations describe wave processes possessing the superposition property. In linear waves, the space-time spectral components of the wave fields propagate without distortion and do not interact with each other.

The linear medium is some idealized model for describing the real environment, and this is not always adequate. The applicability of the linear medium model depends first of all on the magnitude of the ratio of the wave amplitude to the characteristic quantity that determines the properties of the medium. In a linear environment, the ratio of the wave amplitude to the characteristic value of the medium is assumed to be infinitesimal, as a result of which the wave equation becomes linear.

For a finite value of this ratio, it is necessary to take into account *nonlinear terms* in the wave equation. The inclusion of nonlinear terms in the wave equation leads to qualitatively new phenomena. If a monochromatic wave is fed to the input of such a system, then the nonlinearity leads to successive excitation of the time harmonics of the initial wave. The spreading of the frequency spectrum further distorts the shape of the initial sinusoidal wave profile.

In wave systems, the degree of nonlinear interaction is determined both by the considered local nonlinearity and by the ratio of the extent of the interaction region to the wavelength. The extent of the region of effective harmonic interaction largely depends on the dispersion and dissipation of the medium. The energy exchange between the harmonics depends on the phase relationship. In a medium without frequency dispersion, all waves run with the same velocities, and the phase relations remain in the process of propagation between the harmonics. This condition is called the *phase matching* condition. If the attenuation of the waves is small, even minor nonlinear effects can accumulate in proportion to the distance, and the wave will become unstable and breaking over time [Vinogradova, Rudenko, 1979].

In the case of a medium with dispersion, the phase velocities of the waves at different frequencies are different, so that the relations between the phases of the harmonics vary rapidly in space. In case of violation of phase matching nonlinear effects do not accumulate and energy transfer is negligible. Therefore, in the dispersive media there is no noticeable distortion of the shape of the wave profile.

Consider the nonlinear equations, which are often used in fluid dynamics, although they are found in many other areas of modern physics. Taking into account the analogy of nonlinear effects of any nature, one can create a model equation for a one-dimensional wave [Brekhovskikh, 1982], [Gabov, 1988]

$$\frac{\partial u}{\partial t} + Lu = -\varepsilon u \frac{\partial u}{\partial x}. \tag{1.3.1}$$

Here $\varepsilon \ll 1$ is the nonlinearity parameter; L – linear operator, corresponding to a certain dispersion of linear waves.

If $L = c\partial/\partial x$ we obtain equation

$$\frac{\partial u}{\partial t} + c\frac{\partial u}{\partial x} = -\varepsilon u\frac{\partial u}{\partial x}. \tag{1.3.2}$$

This is a nonlinear equation of acoustic type without dispersion and dissipation; its solution is Riemann invariants, leading to different propagation velocities of the compression and extension regions.

If $L = c\frac{\partial}{\partial x} + \beta\frac{\partial^3}{\partial x^3}$ we obtain equation

$$\frac{\partial u}{\partial t} + c\frac{\partial u}{\partial x} + \beta\frac{\partial^3 u}{\partial x^3} = -\varepsilon u\frac{\partial u}{\partial x}. \tag{1.3.3}$$

This equation of *Korteweg and de Vries* for nonlinear media with dispersion describes surface waves in shallow water and is a particular case of a more common shallow water equation

$$\frac{\partial u}{\partial t} + u\frac{\partial u}{\partial x} + g\frac{\partial u}{\partial x} + \frac{1}{3}H\frac{\partial^3 u}{\partial x \partial t^2} = 0. \tag{1.3.4}$$

Equation (1.3.4) is called the *Boussinesq* equation; it describes nonlinear waves of small amplitude, moving both to the left and to the right. For the case of two-dimensional waves in shallow water, the *Kadomtsev-Petviashvili* equation is used

$$\frac{\partial}{\partial x}\left(\frac{\partial u}{\partial t} + \frac{3c}{2H}u\frac{\partial u}{\partial x} + \frac{1}{6}cH^2\frac{\partial^3 u}{\partial x^3}\right) + \frac{1}{2}c\frac{\partial^2 u}{\partial y^2} = 0 \tag{1.3.5}$$

Surface wave profiles in deep water are described by the cubic *Schrödinger* equation

$$i\frac{\partial u}{\partial t} + \frac{\partial^2 u}{\partial x^2} + \nu\,|u|^2\,u = 0. \tag{1.3.6}$$

Schrödinger equation relates to the parabolic type with a space variable.

A description of the surface waves in shallow water with arbitrary dispersion used the *Whitham* equation with integral operator:

$$\frac{\partial u}{\partial t} + u\frac{\partial u}{\partial x} + \int_{-\infty}^{+\infty} K(x-s)u_s(s,t)ds = 0. \tag{1.3.7}$$

To study the effects of viscosity in fluids model equation is the *Burgers* equation:

$$\frac{\partial u}{\partial t}+\varepsilon u\frac{\partial u}{\partial x}=\alpha\frac{\partial^2 u}{\partial x^2}. \qquad (1.3.8)$$

The Burgers equation refers to a parabolic type and is a one-dimensional case of the Navier-Stokes equations. The *Navier-Stokes* equations describe two-dimensional wave processes in a viscous incompressible fluid:

$$\rho\frac{\partial \mathbf{v}}{\partial t}+\rho(\mathbf{v}\nabla)\mathbf{v}=-\nabla p+\eta\Delta\mathbf{v}+(\xi+\frac{\eta}{3})\nabla\nabla\mathbf{v}. \qquad (1.3.9)$$

The process of propagation of a bounded sound beam in a nonlinear medium is described by the two-dimensional *Khokhlov-Zabolotskaya-Kuznetsov* equation:

$$\frac{\partial}{\partial \tau}\left[\frac{\partial p'}{\partial z}-\frac{\varepsilon}{c_0^3\rho_0}p'\frac{\partial p'}{\partial \tau}-\frac{b}{2c_0^3\rho_0}\frac{\partial^2 p'}{\partial t^2}\right]=\frac{c_0}{2}\Delta_\perp p', \qquad (1.3.10)$$

where, $\Delta_\perp=\partial^2/\partial x^2+\partial^2/\partial y^2$ is the Laplacian in transverse coordinates.

In the theory of vibrations and solitons applicable equation to the Klein-Gordon:

$$\frac{\partial^2 u}{\partial t^2}-\frac{\partial^2 u}{\partial x^2}=V'(u), \qquad (1.3.11)$$

here $V'(u)$ is a certain nonlinear function, depending on u. When $V'(u)=\sin u$ the Klein-Gordon equation is transformed into the so-called *sin-Gordon* equation

$$\frac{\partial^2 u}{\partial t^2}-\frac{\partial^2 u}{\partial x^2}=\sin u, \qquad (1.3.12)$$

The sin-Gordon equation is used to describe topological solitons with the geometry of surfaces of negative Gaussian curvature.

The theory of nonlinear waves also uses an equation having a solution of the type of running waves of arbitrary shape.

$$\left(1-\frac{\partial u^2}{\partial t}\right)\frac{\partial^2 u}{\partial x^2}+2\frac{\partial u}{\partial x}\frac{\partial u}{\partial t}\frac{\partial^2 u}{\partial x\partial t}-\left(1+\frac{\partial u^2}{\partial x}\right)\frac{\partial^2 u}{\partial t^2}=0. \qquad (1.3.13)$$

This is the *Born-Infeld* equation; it describes the phase discontinuities in the interaction of two solitons.

1.4 Methods for Solving Nonlinear Equations

The Method of Successive Approximations

There are various approximate analytical methods for solving problems in mathematical physics. The choice of a suitable method for solving equations depends on the nature of the problem under consideration. *Nonlinear equations* can be divided into two classes – algebraic and transcendental. Algebraic equations are equations that contain only algebraic functions (integer, rational, irrational). Equations containing other functions (trigonometric, exponential, logarithmic, and others) are called transcendental [Amosov, 1994], [Shup Terry, 1990].

Methods for solving nonlinear equations can be divided into two groups:

1. *exact methods;*
2. *iterative methods.*

Exact methods allow us to write the roots in the form of a finite relation (formula). As is known, many equations and systems of equations do not have analytic solutions. First of all, this applies to most transcendental equations. In some cases, the equation contains coefficients known only approximately. To solve them, iterative methods are used with a given degree of accuracy.

Solving the equation by the *iterative method* means: to establish whether it has roots, how many roots, and to find the values of the roots with the required accuracy. Approximate values of the roots (initial approximations) can also be known from the physical meaning of the problem, from solving a similar problem with other input data, or can be found *graphically*.

The iterative process consists in the successive refinement of the initial approximation x_0. Each such step is called an iteration. As a result of iterations there is a sequence of approximate values of the root $x_1, x_2, \ldots, x_n$. If these values increase with the number of iterations n approach the true value of the root, then it is said that the iterative process converges. Iterative methods include the method of *successive approximations* (or the simple iteration method).

The essence of the method of successive approximations is as follows. The equation $f(x) = 0$, to be solved is rewritten in the form

$$x = \phi(x) \tag{1.4.1}$$

After this, the initial approximation x_1 is chosen and is substituted into the right-hand side of equation (1.4.1). The obtained value $x_2 = \phi(x_1)$ is taken as the second approximation for the root. If an approximation x_n is found, then the following approximation x_{n+1} is determined by the formula

$$x_{n+1} = \phi(x_n).$$

Suppose that after a few approximations we find that a given degree of accuracy the equality $x_n \approx x_{n+1}$. Since $x_{n+1} = \phi(x_n)$, then this means that the equation is satisfied with a given accuracy $x_n \approx \phi(x_n)$, that is, x_n which is the approximate value of the root of the equation $x = \phi(x)$.

When using the method of successive approximations, it is necessary to find out the following features:

1. if there is equality $\lim_{n\to\infty} x_n = \xi$, then whether the number ξ solution of equation $x = \phi(x)$?
2. always whether the sequence x_1, ..., x_n converges to some number ξ?
3. how fast numbers approach x_1, ..., x_n to the root ξ equations $x = \phi(x)$?

Answering the first question, let's say, that the numbers x_1, ..., x_n ... approaching the number ξ. Consider the equality $x_{n+1} = \phi(x_n)$, Giving an expression of the next approximation in the previous. With increasing n, its left-hand side approaches to ξ, and the right-hand side to $\phi(\xi)$. Therefore in the limit we obtain $\xi = \phi(\xi)$, that is, it ξ is the root of equation $x = \phi(x)$.

The second question is ambiguous; for this we consider the geometric representation of the method of successive approximations [Vilenkin, 1968].

Geometric Interpretation of the Method of Successive Approximations

Finding the root ξ of the equation $x = \phi(x)$ is nothing more than finding the abscissa of the point A of the intersection of the curve $y = \phi(x)$ with the line $y = x$. We construct xOy the graphs of the functions $y = x$ and $y = \phi(x)$. Each real root ξ of equation (1.4.1) is the abscissa of the

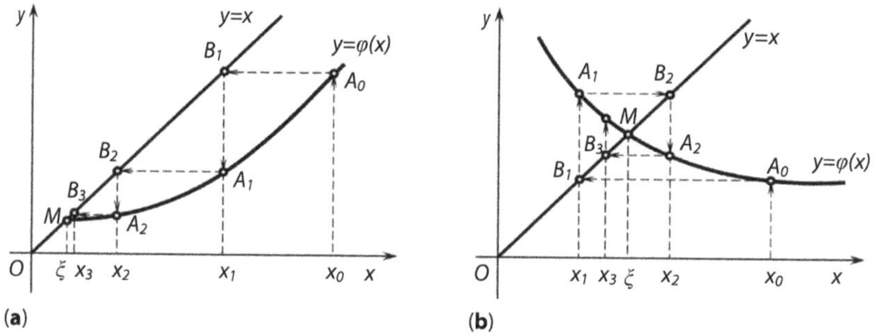

Figure 1.4.1 Convergent iterative processes.

intersection point A of the curve $y=\phi(x)$ with the straight line $y = x$ (Figure 1.4.1a).

Starting from some point $A_0(x_0,\phi(x_0))$, we build a broken line $A_0B_1A_1B_2A_2\ldots$ – ("stairs"), the links of which are alternately parallel to the axis Ox and axis Oy, vertices A_0, A_1, $A_2,\ldots$ on the curve $y=\phi(x)$, and the vertices B_1, B_2, $B_3,\ldots$ – on the line $y = x$. Common abscissae of the points A_1 and B_1, A_2 и B_2,..., obviously represent respectively the successive approximations x_1, x_2,... root ξ.

Another kind of broken line is also possible $A_0B_1A_1B_2A_2\ldots$ – a «spiral» (Figure 1.4.1b). The solution in the form of a «stairs» is obtained if the derivative $\phi'(x)$ is positive, and the solution in the form of a «spiral», if $\phi'(x)$ negative.

Thus, the geometrical meaning of the method of successive approximations is that we move to the desired point of intersection of the curve and the straight line along the broken line. In this case, the vertices of the broken line lie alternately on the curve and on the straight line, and the sides alternately have horizontal and vertical directions (Figure 1.4.1).

In Figure 1.4.1a, b the curve $y=\phi(x)$ in the neighborhood of the root ξ – that is, $|\phi'(x)|<1$, and the iteration process converges. However, if we consider the case when $|\phi'(x)|>1$, then the iteration process can be divergent (Figure 1.4.2). Therefore, for practical application of the method of successive approximations, it is necessary to find out sufficient conditions for the convergence of the iterative process.

1.5 The Basic Laws of Hydrodynamics of an Ideal Fluid

Hydrodynamics studies the motion of liquids and gases; the basic principles of hydrodynamics were established by Euler, Bernoulli and Lagrange.

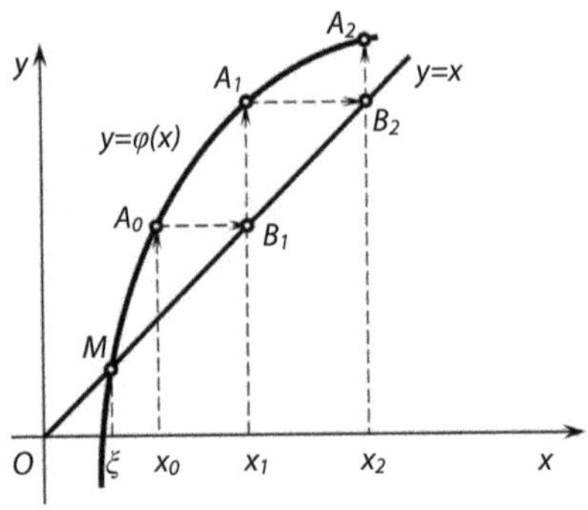

Figure 1.4.2 The divergent iterative process.

To derive the basic laws, we shall first consider the fluid to be ideal, that is, the liquid has no internal friction, and the mechanical energy does not go over into thermal energy. We also neglect heat exchange between different volumes of the liquid. This means that all processes occur at constant entropy, and the stressed state of the liquid is characterized by a single scalar quantity, the pressure p.

The motion of a fluid can be considered definite if all the quantities characterizing the fluid (the particle velocity $\mathbf{v}$, pressure p, density ρ, temperature T etc.), are given as functions of the coordinates and time. This method of specifying fluid motion was proposed by Euler. At the same time, fixing a certain point of space, we follow the change in time of the corresponding quantities at this point. Fixing the moment of time, we find the change of this quantity from point to point. However, no information on what kind of a particle of liquid is at a given point at a given time and how it moves in space, we do not directly have.

Another way of describing the flows, based on the description of the motion of individual liquid particles, was proposed by Lagrange. In the Lagrangian description, attention is fixed to certain particles of the liquid and can be traced, as a change over time of their position, speed, pressure, density, temperature and other quantities in their environment.

These two descriptions of the movement are completely equal, and the choice of one of them in each particular case is dictated only by considerations of convenience. Most instruments measure fluid characteristics at a fixed point, that is, they give Euler information. If you paint (mark) a part of the liquid, then by spreading the paint, Lagrangian motion information is obtained. Therefore, the Lagrange method is simpler; it describes the diffusion process associated with the movement of the particles.

The System of Equations of Hydrodynamics

The continuity equation. One of the fundamental equations of hydrodynamics is the equation of continuity, or the law of conservation of matter. It shows that the mass of a liquid in a volume that covers the same particles all the time is preserved. Mathematically, this can be written as follows [Brekhovskikh, 1982]:

$$\frac{d}{dt}\int_V \rho dV = 0 \tag{1.5.1}$$

or in partial derivatives

$$\int_V \left(\frac{\partial \rho}{\partial t} + \mathbf{v}\nabla\rho + \rho\nabla\mathbf{v} \right) dV = 0 \tag{1.5.2}$$

whence, by virtue of the arbitrariness of the volume V, equating the integrand with zero, we obtain the required equation of continuity

$$\frac{\partial \rho}{\partial t} + \mathbf{v}\nabla\rho + \rho\nabla\mathbf{v} = 0 \text{ or } \frac{\partial \rho}{\partial t} + \nabla(\rho\mathbf{v}) = 0, \text{ or } \frac{d\rho}{dt} + \rho\nabla\mathbf{v} = 0 \tag{1.5.3}$$

The integral form of the continuity equation (1.5.2) has a simple physical meaning (with the Ostrogradskii-Gauss theorem)

$$\frac{\partial}{\partial t}\int_V \rho dV = -\int_S \rho\mathbf{v}\mathbf{n} dS \tag{1.5.4}$$

– the speed change of the mass of the liquid inside a fixed volume V is equal to the mass of the liquid (with the opposite sign) flowing out of the volume through its surface per unit time S.

The Euler equation. We now turn to the derivation of the equation of fluid motion, also called the Euler equation. For this it is sufficient to apply Newton's second law: the time derivative of the momentum of a certain volume of liquid is equal to the sum of the forces acting on this volume:

$$\frac{d}{dt}\int_V \rho\mathbf{v} dV = \mathbf{F} + \mathbf{F}_S \tag{1.5.5}$$

where

$$\mathbf{F}=\int_V \rho\mathbf{f}dV$$

is the external volume force (**f** – is the force per unit mass); $\mathbf{F}_s$ – is the force acting on the volume *V* from the environment side through the bounding surface *S*,

$$\mathbf{F}_S=-\int_S p\mathbf{n}dS\,,$$

where **n** – is the outer normal to *S*.

Using the arbitrariness of the volume *V*, we obtain the desired Euler equation

$$\frac{d\mathbf{v}}{dt}=\frac{\partial\mathbf{v}}{\partial t}+(\mathbf{v}\nabla)\mathbf{v}=-\frac{\nabla p}{\rho}+\mathbf{f} \tag{1.5.6}$$

Completeness of the system of equations. Equation (1.5.6) together with the continuity equation (1.5.3) is four scalar equations for five scalar quantities (density ρ, pressure p and three components of the velocity vector ϑ_{i}). Consequently, the system of equations is not yet closed. The thermodynamic equation of state connecting the three quantities: pressure p, density ρ and entropy s

$$p=p(\rho,s), \tag{1.5.7}$$

and the equation for entropy. As we noted, the entropy of a given liquid particle remains constant, i.e.

$$\frac{ds}{dt}=\frac{\partial s}{\partial t}+\mathbf{v}\nabla s=0 \tag{1.5.8}$$

Entropy s can be excluded from the equations, for this we need to take from (1.5.7) take the time derivative and denote by

$$c^2=(\partial p/\partial\rho)_s \tag{1.5.9}$$

As a result, we obtain

$$\frac{dp}{dt}=c^2\frac{d\rho}{dt} \tag{1.5.10}$$

where c^2 should already be considered a given functions p and ρ, and is the speed of sound in the medium.

If the liquid is barotropic, i.e. the pressure depends only on the density

$$p = p(\rho) \tag{1.5.11}$$

To then the last equation will be the closure for the system (1.5.3) and (1.5.6). The form of the function $p(\rho)$ depends on the properties of the liquid or gas under consideration.

A complete closed system of equations (1.5.3), (1.5.6) and (1.5.10) is called the system of hydrodynamics equations for an ideal fluid:

$$\frac{\partial \rho}{\partial t} + \mathbf{v}\nabla\rho + \rho\nabla\mathbf{v} = 0 \quad \text{– the continuity equation,}$$

$$\frac{d\mathbf{v}}{dt} = \frac{\partial \mathbf{v}}{\partial t} + (\mathbf{v}\nabla)\mathbf{v} = -\frac{\nabla p}{\rho} + \mathbf{f} \quad \text{– the Euler equation (motion),}$$

$$\frac{dp}{dt} = c^2 \frac{d\rho}{dt} \quad \text{– equation of state.}$$

To solve specific problems, the corresponding boundary conditions must be added to the equations. An example of the force **f**, entering the Euler equation (1.5.6) is the force of gravity. In this case **f** is equal to the vector $g(g = |g| = 9{,}81\ m/s^2)$, directed to the center of the Earth. The equations can also include the forces of attraction of the moon and the sun, which are taken into account in the theory of the sea tides. In the dynamics of the ocean and the atmosphere, the non-inertiality of the reference system associated with the rotating Earth is essential. In this case inertia forces must be introduced into the system of equations of hydrodynamics:

$$\text{centrifugal } \mathbf{f}_{\text{ctr}} = \Omega^2 r = \nabla(\Omega^2 r^2 / 2),$$

where Ω – is the Earth's rotation frequency, and the Coriolis force $\mathbf{f}_k = -2\Omega \times \mathbf{v}$. We note that all the forces considered by us, other than the Coriolis force, are potential, that is, representable in the form of a gradient from the potential function: $\mathbf{f} = -\nabla u$.

In connection with the assumption of incompressibility of a fluid, the density of each particle must remain constant:

$$\frac{d\rho}{dt} = \frac{\partial \rho}{\partial t} + \mathbf{v}\nabla\rho = 0.$$

It does not follow from the equation of state (1.5.10) that $dp/dt = 0$, since in an incompressible fluid the speed of sound c is infinitely large. Thus, in an incompressible fluid equation (1.5.10) is replaced by $d\rho/dt = 0$ and, taking into account the continuity equation, we have

$$\frac{\partial \rho}{\partial t} + \mathbf{v}\nabla\rho = 0, \quad \nabla \mathbf{v} = 0,$$

thus the system of hydrodynamic equations remains closed.

1.6 Linear Equations of Hydrodynamic Waves

Due to wave movements carried out in a fluid energy transfer over large distances without mass transfer. The various forces acting on liquid particles during their motion lead to different wave motions. Let us consider the main types of waves in a liquid in the so-called linear approximation, when the waves propagate independently of each other. Gravity surface waves and internal waves in the thickness of a liquid are related to waves due to the action of gravity. As a result of the compressibility of the medium, acoustic waves propagate in it.

Linearization of the equations of hydrodynamics. When studying wave processes in a liquid, we must start from the basic equations of hydrodynamics ((1.5.3), (1.5.6) and (1.5.10)) [Brekhovskikh, 1982]:

$$\begin{aligned}
&\frac{\partial \mathbf{v}}{\partial t} + (\mathbf{v}\nabla)\mathbf{v} = -\frac{\nabla p}{\rho} - g\nabla z - 2\Omega \times \mathbf{v}, \\
&\frac{\partial \rho}{\partial t} + \nabla(\rho \mathbf{v}) = 0, \\
&\frac{dp}{dt} = c^2 \frac{d\rho}{dt}, \qquad c^2 = \left(\frac{\partial p}{\partial \rho}\right)_s
\end{aligned} \tag{1.6.1}$$

Here, in Euler's equation, we included the force of gravity directed vertically downward, and the Coriolis force $-2\Omega \times \mathbf{v}$, arising in a fluid rotating with frequency Ω (for example, on a rotating Earth). Let us consider the theory of wave propagation in an ideal fluid, without touching on the questions of their generation.

An important circumstance in the study of wave motions in a liquid is the nonlinearity of the hydrodynamic equations (1.6.1), so the exact theory of waves in a liquid is a nonlinear theory. This will lead to the influence of some wave processes on the other (the interaction of waves) and to a significant complication of the process of propagation of waves of each individual species. However, if the perturbations of the fluid caused by the waves are small, then the hydrodynamic equations can be linearized relatively to these perturbations.

The theory of waves in a liquid becomes linear and the superposition principle comes into force. In the hydrodynamic equations, the value of the velocity ϑ will be assumed to be a small value, and the perturbed pressure p and density ρ values differ little from their equilibrium values p_0 and ρ_0 accordingly. After this, putting $p = p_0 + p'$, $\rho = \rho_0 + \rho'$, we transform equations (1.6.1), preserving only the linear in $\mathbf{v} \equiv \{u, \upsilon, w\}$, p' and ρ' the terms (the linearization of the equations). For the equilibrium state we have the equations of hydrostatics:

$$\mathbf{v}_0 = 0, \qquad \frac{dp_0}{dz} = -g\rho_0(z) \tag{1.6.2}$$

When linearization the nonlinear term $(\mathbf{v}\nabla)\mathbf{v}$ in the Euler equation falls, a member of $\nabla p/\rho$ the following transformation:

$$\frac{\nabla p}{\rho} = \frac{\nabla p_0 + \nabla p'}{\rho_0(1+\rho'/\rho_0)} \approx \frac{\nabla p_0}{\rho_0} + \frac{\nabla p'}{\rho_0} - \frac{\nabla p_0 \rho'}{\rho_0 \rho_0}$$

Similarly, a nonlinear term in the continuity equation $\nabla(\rho\mathbf{v}) = \nabla[(\rho_0 + \rho')\mathbf{v}] \approx (d\rho_0/dz)w + \rho_0\nabla\mathbf{v}$
and in the equation of state

$$\frac{dp}{dt} \approx \frac{\partial p'}{\partial t} + \frac{\partial p_0}{\partial z}w \approx c^2\left(\frac{\partial \rho'}{\partial t} + \frac{d\rho_0}{dz}\right)w.$$

If we now recall that $\nabla p_0 = (dp_0/dz)\nabla z$ and also take into account the hydrostatics equation (1.6.2), then it is easy to write out the linear system of hydrodynamic equations (the strokes p' and ρ' omitted):

$$\frac{\partial \mathbf{v}}{\partial t} + \frac{\nabla p}{\rho_0} + g\frac{\rho}{\rho_0}\nabla z + 2\Omega \times \mathbf{v} = 0, \tag{1.6.3}$$

$$\frac{\partial \rho}{\partial t} + \frac{\nabla \rho_0}{z}w + \rho_0\nabla\mathbf{v} = 0,$$

$$\frac{\partial \rho}{\partial t} = \frac{1}{c^2}\frac{\partial p}{\partial t} + \rho_0 \frac{N^2(z)}{g} w,$$

where $N^2(z) = -g(\rho_0^{-1} d\rho_0 / dz + g / c^2)$ — Vaisala frequency.

Linear boundary conditions. For free waves, the forces that arise at the boundaries of a region occupied by a liquid can act as "returning" forces. The most important from this point of view is the free surface. Suppose that in a state of equilibrium the surface of a liquid is plane. When removing it from this position, there are two kinds of forces that tend to return the surface to its previous state: gravity and surface tension. As a result, a perturbation appears on the surface of the liquid, propagating in all directions in the form of a wave.

The forces arising on the surface of the liquid must enter the boundary conditions for equations (1.6.3). We get them under the assumption that in a state of equilibrium the free surface is flat ($z = 0$). Let the rise of the perturbed surface be described by expression $\zeta = \zeta(x, y, t)$. Surface elevation speed $d\zeta / dt$ should coincide with the vertical particle velocity $w(x, y, z, t)$, of the medium that can be written as

$$w\Big|_{z=\zeta(x,y,t)} - \frac{d\zeta(x, y, t)}{dt} = 0 \qquad (1.6.4)$$

This is the so-called kinematic boundary condition on the free surface of a liquid.

In the second, dynamic condition on the free surface of the fluid, the pressure jump on both sides of the boundary due to the surface tension forces is taken into account. If the pressure in the liquid $p_0(z) + p$, and the pressure above the surface p_a, then by the well-known Laplace formula

$$p_0(\zeta) + p\Big|_{z=\zeta} - p_a = -\sigma R^{-1}, \qquad (1.6.5)$$

where σ – surface tension coefficient; R^{-1} – the sum of the principal curvatures of the surface ζ, equal to the surface $z = \zeta(x, y, t)$

$$R^{-1} = \nabla(\nabla\zeta / \sqrt{1 + (\nabla\zeta)^2}) \qquad (1.6.6)$$

The boundary conditions (1.6.4) and (1.6.5) also turned out to be non-linear. We linearize them, assuming that the perturbation of the surface ζ is small. In this case, applying the formula

$$\frac{d}{dt} = \frac{\partial}{\partial t} + (\mathbf{v}\nabla)$$

for the total time derivative, from (1.6.4) we obtain

$$w\Big|_{z=0} + \frac{\partial w}{\partial z}\Big|_{z=0}\zeta + \ldots - \frac{\partial \zeta}{\partial t} - \mathbf{v}\nabla\zeta\Big|_{z=0} - \ldots \approx w\Big|_{z=0} - \frac{\partial \zeta}{\partial t} = 0.$$

Similarly, from (1.6.5) taking into account (1.6.6) and $p_0(0) = p_a$ have a

$$\frac{dp_0}{dz}\Bigg|_{z=0}\zeta + p\Big|_{z=0} + \ldots = -\sigma\Delta\_\zeta + \ldots,$$

where $\Delta\_ = \partial^2/\partial x^2 + \partial^2/\partial y^2$ — Laplace operator on the horizontal variable. As a result, limiting ourselves to linear terms only, taking into account the hydrostatic equation (1.6.2), we obtain linear boundary conditions on the unperturbed free surface $z = 0$:

$$w\Big|_{z=0} = \frac{\partial \zeta}{\partial t}, \quad p\Big|_{z=0} = g\rho_0(0)\zeta - \sigma\Delta\_\zeta. \tag{1.6.7}$$

Linear equations for hydrodynamic waves. In the basic equations (1.6.3) for waves in a liquid, the change in its density due to the compressibility is taken into account, which is described by the first two terms in the equation of state:

$$\frac{\partial \rho}{\partial t} = \frac{1}{c^2}\frac{\partial p}{\partial t} \tag{1.6.8}$$

The most important consequence of the compressibility of a liquid is acoustic waves, for surface and internal waves the compressibility is demonstrated only in the form of some small corrections that we neglect.

Formally, the transition to an incompressible fluid can be carried out by requiring a constant density in the liquid particle:

$$d\rho/dt = 0 \tag{1.6.9}$$

Here, directly from the exact continuity equation (1.5.3) it follows that when the fluid moves

$$\nabla \mathbf{v} = 0 \tag{1.6.10}$$

On the other hand, from the exact equation of state (15.10), under the condition (1.6.9), we have $c^{-2} dp / dt = 0$. But in the general case $dp / dt \neq 0$, for example, because of a change in the hydrostatic pressure, so in an incompressible fluid, we must consider the speed of sound $c = \infty$. This result reflects the fact that in an incompressible fluid, elastic perturbations propagate instantaneously. Now it is no longer difficult to obtain from (1.6.3) the *linear equations* of the hydrodynamics of an incompressible fluid:

$$\begin{aligned}
&\frac{\partial \mathbf{v}}{\partial t} + \frac{\nabla p}{\rho_0} + g\frac{\rho}{\rho_0}\nabla z + 2\Omega \times \mathbf{v} = 0, \\
&\nabla \mathbf{v} = 0, \\
&\frac{\partial \rho}{\partial t} - \rho_0 \frac{N^2}{g} w = 0,
\end{aligned} \tag{1.6.11}$$

where Vaisala frequency $N^2(z) = -g\rho_0^{-1} d\rho_0 / dz$. The boundary conditions (1.6.7), which do not contain the velocity of sound *c*, remain unchanged. The condition for the applicability of the approximation of an incompressible fluid ($c = \infty$) is the requirement that the propagation velocity of the perturbations under consideration be significantly less than the speed of sound.

Conclusions

The first chapter described the features of mathematical physics. A classification of the linear differential equations of the second order, which are given many mathematical physics tasks, is provided. Additional initial and boundary conditions for these tasks are provided.

The tasks of hydrodynamics, which result in nonlinear hydrodynamic equations, are described, and the often-used nonlinear equations are provided. Methods for solving nonlinear equations, the method of successive approximations, the conditions for the applicability of this method, and its geometric interpretation are considered. The main ways of describing the fluid motion, the complete system of hydrodynamic equations, are shown. Linearization of the equations of hydrodynamics have been carried out.

The results of the studies in this chapter have been published in the following main works: [Abbasov, 2007 Fizmatlit], [Abbasov, 2010 Fizmatlit], [Abbasov, 2001], [Abbasov, 2003], [Abbasov, 2004].

2

Modeling of Wave Phenomena on the Shallow Water Surface

2.1 Waves on the Sea Surface

The most common type of fluid movement is waves. They can occur for a variety of reasons. The waves on the sea surface are subject to wind effects and are preserved after it for some time in the form of *swell*. Wave movements, covering the entire sea water and long-wave class, are brought on by the moon and the sun *(tidal waves)*, underwater earthquakes *(tsunami)* or cyclone activity [Monin, 1985].

Free surface waves exist in the absence of wind on the sea surface. Like wind waves, free waves can also be accidental. Free waves on the surface of the fluid are created by two factors—gravity and surface tension. Both factors operate simultaneously, but if the waves are long enough (with length of $\lambda \geq 10$ cm), the gravity prevails. Such waves are called *gravity*; for them, the surface tension is insignificant. Short waves (with a length of $\lambda \leq 0.4$ cm) are due only to surface tension. Such waves are called *capillary*; they have no significant gravity. Waves whose lengths are in the intermediate range are called *gravitational-capillary*; they have both a significant force of gravity and a surface tension. Waves from capillary, capillary-gravitational wavelength ranges are often called *ripples*.

Surface gravity waves are characterized primarily by their λ and amplitudes a, and depend on the depth of fluid H. When the water depth is more than half the wavelength, it is said that waves on deep water "don't feel" the influence of the floor. The lower H/λ corresponds to the waves on the shallow water, and in the limits $H/\lambda \to 0$ There may be a private wave (soliton) consisting of only one back. In deep water, in the wave of the low amplitude, individual water particles hardly move with the wave, but move in orbits close to the circular, whose radii are exponentially decreasing with the depth. In shallow water and near the bottom, these orbits turn into ellipses whose vertical axis is less than horizontal.

Low amplitude waves are an important private case, when $a/\lambda << 1$. In this case, the surface of the water is presented as a superposition of simple harmonic waves and, in the simplest case, in the form of a single harmonic wave, when the surface of the water is the type of one "running" cosine. The low amplitude wave theory is called linear, and it explains many important surface waves properties. Under the influence of nonlinear effects in the gravity waves their crest became sharper, thoughts became flat in comparison with cosine wave of low amplitude.

For wind speeds of less than 1 m/s, capillary waves will be formed on the calm surface of the water. They occur very quickly in the gusts of the wind, creating stains of the ripples on the surface, but they also quickly decay by molecular viscosity. With the wind rise, larger, gravity wind waves appear. In real-world uses, the ocean surface has various types of waves – from capillary waves with the length shorter than a centimeter to gravity wave swells with the length of hundreds of meters. The maximum wave height depends on the wind speed, duration and acceleration, the distance at which the wind continues to impact the generated and spreading wave. The longer the wind or the acceleration impact, the higher the waves come.

Wind waves on the ocean surface have a random nature. Therefore, statistical characteristics are used to describe them. The spectral waves are commonly used, which determine the distribution of their energy at different frequency intervals. Spectral wind waves are quite narrow in both frequencies and distribution directions. Most of the energy is concentrated in frequency interval with the length of one octave. The phase velocity of the energy-carrying components of wind waving is close to the average wind speed. In a typical wind velocity range of $10 \div 30$ m/s, the range of wavelengths of energy-carrying components is $60 \div 600$ m, the range of periods is $5 \div 20$ s and the range of heights is $2 \div 20$ m.

In the Atlantic, the waves rarely exceed 12 *m*; in the Pacific Ocean they can reach 15 *m*, slightly larger in the stormy areas of Antarctica. When the wind comes off or waves leave the area of the storm, they become more restrictive and their altitude decreases. The waves turn into swell and continue to spread over long distances (about a thousand miles) until the shore is reached.

When the surface gravity waves came to the coast or shallow water, their characteristics change connected with shoaling. The waves are transformed when they start to "feel" the floor. The nature of the transformation depends on the topography of the floor and the configuration of the coastline. If, when the waves are distributed, their crests are not parallel to the isobath (lines of equal depth), there is a *refraction* – wave rays are no longer straight lines (as in deep water), they curve and seek to become normal to the coast, and the fronts (the crest) are parallel.

The spread of the fronts (the crest) is due to the fact that the front areas closer to the coast (at a lower depth) are moving at a slower speed than the areas at greater depth. In the refraction of waves over the uneven bottom, their amplitudes and lengths change, and the frequencies remain almost constant.

The spectrum of waves is subject to significant changes. Wave rays are also twisted when *diffraction* waves around obstacles. In this case, the ray curvature is no longer associated with the depth changes.

The waves that fall in the shallow water of the cape are moving slower than the floor of the inlet inside the bay. As a result, the wave front is bent and the wave rays converge near cape and diverge in the bays. This results in a concentration of wave energy at the cape and its scattering in the bays.

When the wave is approaching the coast, its back moves faster than low, and the orbital velocity of the water particles on the back of the wave may exceed its phase velocity. At the moment when "the back is kicking the low", the front slope of the wave becomes plumb, and the wave falls (Figure 2.1.1). The formation of falling waves in the coastal zone is called *surf*.

With sufficiently large slopes of the floor (>0.03), the coast has falling waves in the form of *"diver breaker"*, when the water on the back forms a "twist" falling on the front of the wave. Smaller slopes of the bottom of the

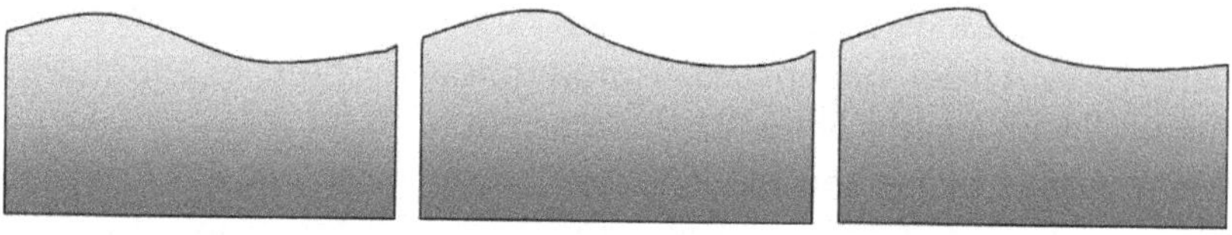

Figure 2.1.1 Transformation of the surface wave profile.

falling waves may follow the type of "*sliding breaker*", when the water flows through its front slope in the direction of the inclined plane. The sliding breaker in the ground swell area is used by surfing enthusiasts. A solitary wave consisting of one back is also possible in the coastal zone.

A particular type of waves in the sea is *rip*. When rip is on the sea surface, it combines wave and vortex movements. Rips lack visible movement; they resemble very irregular standing waves. The wave height at rip reaches 4–5 *m*. The rip produces a distinctive sizzling sound.

The gravity forces of the moon and the sun create tides in the ocean, causing fluctuations in its level and having the character of long waves that are much larger than the ocean depth. When a tidal wave is propagated upstream of certain rivers drainage into the ocean, there is a phenomenon called *tidal surge*. The ideal tidal surge is a "breaking wave" of the water level that spreads through undisturbed water.

When vessels move, the surface of the water resents, and a trail is formed. Usually the vessel's trail consists of *wave trail* and *turbulence trail*. A wave track is a regular wave system consisting of two different wave types, longitudinal and transverse. Longitudinal waves are distributed along the two straight-side lines of the mark that intersect the head of the vessel at an angle of $2\theta_0$, which is not dependent on the speed of the vessel and, according to theory, is about 39°. The transverse waves spread over the stern of the vessel in the direction of its movement and at its speed.

In the oceans, *internal waves* can occur in layers of different densities. If the lighter water layer is located above the layer of the heavier water, there is a dividing line which is the same as the boundary between the atmosphere and the ocean. Ocean waters usually have a stable stratification. Therefore, internal waves are observed everywhere in all parts of the world's oceans.

2.2 Review of Research on Surface Gravity Waves

Waves on the water surface are one of the most common types of wave movement in nature, available for visual observation. The surface of the ocean is almost always covered with waves, and simultaneously there are a variety of waves from the capillary with lengths of the order of millimeters to the swell with lengths of hundreds of meters.

Many results of the theory of surface waves have long been related to classical hydrodynamics. But we cannot say that the problem of surface waves is now finally solved. In recent years, there have been many important results, both in experiments and theory, especially for nonlinear waves.

The problem of adaptation of liquids to the action of gravity in qualitative form can be found even in the work of Archimedes (287-212 B.C.) "On floating bodies" [Gill, 1986]. In a quantitative form, this task required first the discovery of the laws of motion and the development of calculi necessary for the application of these laws. These discoveries were presented in the work of I. Newton "Mathematical Principles of Natural Philosophy" [Newton, 1686] (printed with the help of his friend E. Halley). By this time the laws of hydrostatics and the nature of the pressure forces were described in detail in B. Pascal's works.

When the material element moves, its mass is constant, but its volume can vary. Therefore, its density can vary, and this change depends on the field of fluid velocities. The equation relating the rate of change in the density to the velocity field of the fluid is called the mass conservation equation. There are two equivalent ways of writing this equation in accordance with various methods of its derivation. In the first method, applied by Euler [1755], the work on the equations of motion deals with changes occurring with a material element in the process of its motion. He composed the equation of continuity and the momentum equation.

$$\frac{d\mathbf{v}}{dt} = \frac{\partial \mathbf{v}}{\partial t} + (\mathbf{v}\nabla)\mathbf{v} = -\frac{\nabla p}{\rho} + \mathbf{f}, \tag{2.2.1}$$

where $\mathbf{v}$ – velocity vector, p – pressure, ρ – density of the medium, $\mathbf{f}$ – force per unit mass.

In the second method, proposed by J. Lagrange [1781], changes occurring in a fixed volume element are considered.

Among the first problems considered with the aid of the equations of motion was the problem of the response of the ocean and the atmosphere to gravity. In his paper "On Waves", Laplace (1778) considers the problem of the equilibrium state of a homogeneous liquid of constant depth, in which the free surface initially had a small deviation. In this paper, P. Laplace obtained the equation of fluid motion on a rotating sphere under the action of tidal forces:

$$\nabla^2 p' = \frac{\partial^2 p'}{\partial x^2} + \frac{\partial^2 p'}{\partial y^2} + \frac{\partial^2 p'}{\partial z^2} = 0, \tag{2.2.2}$$

where $p = -\rho g z + p'$ – perturbation of equilibrium pressure, g – gravity constant.

P. Laplace deduced the dispersion relation, which shows that the perturbations are removed from the deflection region with a velocity that depends on the curvature of the surface of the liquid [Gill, 1986].

$$\omega^2 = gk\,\mathrm{th}\,kH\,, \tag{2.2.3}$$

where k – wave number, H – depth of fluid.

The consequence of the dispersion relation is the dependence of the phase velocity on the wave number. Therefore, waves with different lengths emanating from the same point will move with different velocities. Consequently, the waves will disperse or scatter. This phenomenon is called dispersion. The dispersion effect is pronounced in ocean waves generated by a remote storm. Since long waves move faster, they come first and can outpace one or two days of shorter waves coming from the same storm.

In the work [Lagrange, 1781] Lagrange introduced shallow water equations output (at $kH << 1$). The wave equation in the case of constant depth has the form

$$\frac{\partial^2\zeta}{\partial t^2} = c^2\left(\frac{\partial^2\zeta}{\partial x^2}+\frac{\partial^2\zeta}{\partial y^2}\right) = c^2\nabla^2\zeta\,, \tag{2.2.4}$$

where $c_2 = gH$ – velocity of propagation, ζ – surface elevation.

This equation has a solution of the form

$$\zeta = \zeta_0\cos(kx + ly - \omega t)\,,\ k = (k,l) \text{ – is wave vector.}$$

As noted by J. Lagrange, the wave equation looks just like the equation for sound propagation, and there is a complete analogy between shallow-water waves of small amplitude and two-dimensional low-amplitude sound waves.

The theory of shallow water after J. Lagrange was supplemented by G. Airy [Airy, 1845]. He first formulated the limiting equations for the analysis of very long waves of finite amplitude in shallow water. But the dispersion effects were not taken into account in the Airy equations. G. Airy also investigated the propagation of waves in a channel. If the amplitude of the wave is comparable to the average depth, then in the propagation in a channel of a rectangular cross section, its shape changes.

G. Airy, using the *method of successive approximations*, found that in a progressive wave various parts move with different velocities: in the direction of the positive axis x

$$A = A_0\sqrt{1+\frac{\eta}{H}} + u\,, \tag{2.2.5}$$

in the direction of the negative x axis propagates at a speed of

$$A = A_0\sqrt{1+\frac{\eta}{H}} - u. \tag{2.2.6}$$

Since the speed of the wave grows along with the rise, it turns out that in the progressive wave system the slope on the front side becomes more steep, and on the back side it is all the more gently sloping. Here, one can achieve a state in which vertical acceleration cannot be neglected. Therefore, this theory becomes inapplicable, and observations show that the crests tend to curl into a swirl and then decay.

By a similar method, Riemann (1892), in presenting the acoustic problem, established that the velocity of propagation of the vertices and valleys of the wave differ (Riemann invariants). Powerful perturbations move with greater speed, catching up with the weak. This leads to a change in the shape of the wave, to the steepening of the leading edge of the perturbation.

In 1816, for the solution of the problem of deep water waves created by local perturbation, the French Academy announced a prize [Yuen, 1987]. This problem was solved in the same year independently by A. Cauchy (Cauchy, 1815) and S. Poisson (1816).

Wave motions with insignificant velocities were considered, and the deviation of the liquid surface is also small. Later, this theory became known as the *theory of infinitesimal waves*. The simplification of the wave problem was proposed by A. Cauchy. He obtained the boundary condition for the velocity potential $\phi(x,y,z)$:

$$\left(\frac{\partial^2\phi}{\partial 5^2} + g\frac{\partial\phi}{\partial z}\right)_{z=0} = 0 \tag{2.2.7}$$

The velocity potential function $\phi(x,y,z)$ must satisfy the flow condition of the walls of the vessel and the resulting boundary condition. For this theory, the boundary condition must be satisfied on the unperturbed waves of the horizontal plane $z = 0$. For a known function $\phi(x,y,z)$, the equation of an open surface is differentiated by the boundary condition.

However, the system of exact equations describing the shape of the wave motion for an infinite depth of liquid was established as early as 1802 by the Czech scientist F. Gerstner (1809), and later independently by Rankin (1863). A feature of this wave motion is its swirl character.

F. Gerstner presupposed in advance that particles of a liquid in a wave describe circles whose radii decrease with depth. This assumption was based on the results of direct observations of the behavior of floats in the water. His equation of motion had a solution in x and z in the form of the Trochoid equation [Lamb, 1947]:

$$x = \frac{\theta}{k} + a\sin\theta,\ z = -a\cos\theta. \tag{2.2.8}$$

Trochoid is obtained as a result of the motion of a point located on the radius between the center and the rim of the wheel rolling along a horizontal straight line located at the height of the unperturbed water surface (Figure 2.2.1).

Therefore, such a wave is called *trochoidal*, the crests of such a wave lie above the calm surface by an amount $\xi_{crest} = a + \pi/\lambda a^2$, and the trough $\xi_{trough} = a - \pi/\lambda a^2$.

The fluid particles in them move along circles whose radius decreases exponentially with depth. In the extreme case, the wave takes the form of a cycloid, the steepness of such a wave is $2a/\lambda \approx 0,32$, and in nature usually the steepness does not exceed 0,15.

However, the trochoidal waves are accompanied by a swirl motion of the liquid and this is an essential shortcoming of this theory. High waves with sharp crests can not continue its run, because their crests overturned and destroyed, and the waves are reduced in height.

The first steps towards the general theory of finite-amplitude waves accompanied by potential fluid motion were made in 1847 by J. G. Stokes [Stokes, 1847] in his paper "On the theory of oscillating waves." Solving the equation of motion by the method of successive approximations, he obtained formulas for calculating the profile of the wave and elucidated the features of motion.

The Stokes wave equation has the following form [Lamb, 1947]

$$\xi = a\cos(kx - \omega t) - \frac{1}{2}ka^2\cos 2(kx - \omega t) + \frac{3}{8}k^2a^3\cos 3(kx - \omega t) \tag{2.2.9}$$

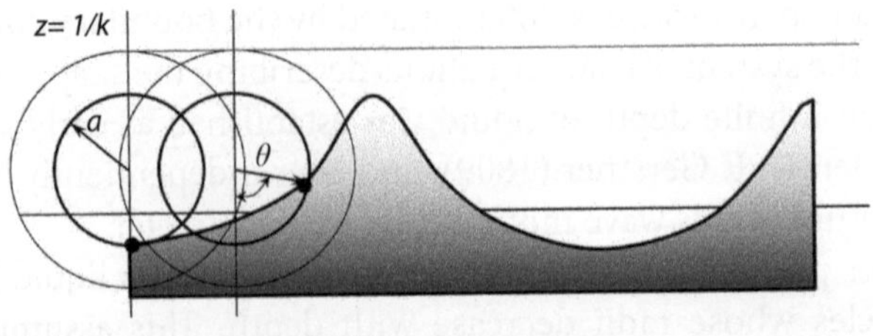

Figure 2.2.1 The trochoidal wave.

Stokes wave is symmetrical about the vertical lines passing through the crests or troughs, but is not symmetrical about the plane of the undisturbed level. The crest of such a wave lies above this plane by an amount

$$\xi_{crest} = a\left(1 + \frac{1}{2}ka + \frac{3}{8}k^2a^2\right),$$

and the troughs is lower by an amount

$$\xi_{trough} = a\left(1 - \frac{1}{2}ka + \frac{3}{8}k^2a^2\right).$$

At the same time, troughs have a more flat outline than crests. With increasing altitude for a given length, the crests of the waves become more sharpened and approach the limiting form at the corner point 120^0.

J. Michell [Michell, 1893] numerically found a periodic wave of maximum height for water with infinite depth and found that the limiting height is reached at $2a/\lambda = 0{,}142$.

The speed of propagation of the Stokes wave depends not only on the length, but also on the amplitude

$$c = \sqrt{\frac{g}{k}\left(1 + k^2a^2\right)} \tag{2.2.10}$$

Figure 2.2.2 shows the profile of the Stokes wave at $ka = 1/2$ or $a/\lambda = 0{,}00796$ [Lamb, 1947]. The profile of the wave (approaching the trochoid) has a contour, sharper on the crests and gently sloping in the troughs.

The particle trajectories (Figure 2.2.3) in the wave motion turn out to be unclosed because of the presence of a wave flow with velocity

$$v_{wave} = a^2k^2\sqrt{\frac{g}{k}}e^{2kz}. \tag{2.2.11}$$

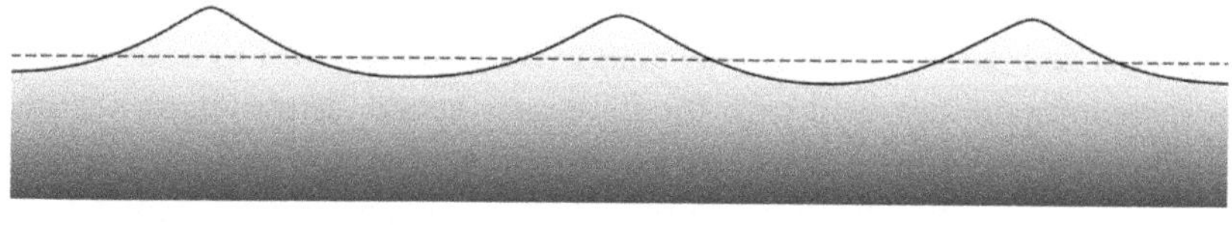

Figure 2.2.2 Stokes wave profile.

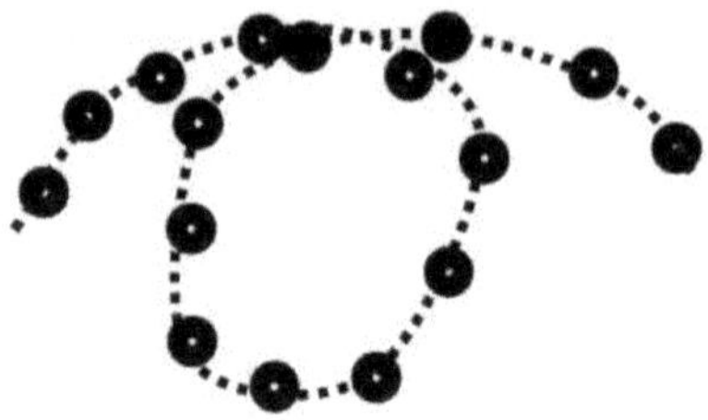

Figure 2.2.3 Trajectory of particle motion in wave.

For waves having ultimate shape and propagating in calm water, the particles on the crests are moving with the same speed of advance as the wave itself. Another interesting property of these wave forms is established that they have relative to the undisturbed water a certain amount of movement in the direction of wave propagation.

The proofs of the convergence of the Stokes expansions were given by T. Levi-Civita in 1925 [Levi-Civita, 1925] and D. Struik in 1926 [Struik, 1926]. Although for the first time a rigorous proof of the existence of periodic steady-state waves of finite amplitude on the surface of a fluid of infinite and finite depth was given in 1921 by A. I. Nekrasov [Nekrasov, 1921]. However, the convergence of the Stokes expansions does not mean stability for the Stokes waves, either, instability exists in deep water.

The evidence of the existence of periodic steady waves in an inhomogeneous incompressible fluid was given by M. Dubreil-Jacotin in 1937 [Dubreil-Jacotin, 1937]. The shape of the streamline of the steady flow of an inhomogeneous fluid was determined, and an equation describing the steady-state waves was obtained.

In 1834, a curious wave on the water was observed by the Scottish scientist John Scott Russell [Russell, 1844]. He studied the capacity of the Union Canal near Edinburgh, and watched the barge move, which was pulled by a pair of horses. Suddenly the barge stopped, but the mass of water that the barge had set in motion did not stop, but gathered at the bow of the ship, and then broke away from it. Further, this mass of water rolled along the channel with great speed in the form of a secluded elevation, without changing its shape and not reducing the speed. Its height was about 40 cm, length about 9 m, speed about 10 km / h (nine miles per hour) [Zarembo, 1984].

He watched this wave on horseback, which allowed him to accompany this wave long enough and with the necessary speed. Throughout his life, Russell repeatedly returned to observe this wave, because he believed that the "*solitary wave*" that he discovered and named plays an important role in many phenomena in nature. He established some properties of this wave.

First, he noticed that it moves at a constant speed and without changing its shape. Secondly, he found the dependence of the velocity c of this wave on the depth of the channel *H* and the height of the wave *a*:

$$c = \sqrt{g(a+H)}\,. \tag{2.2.12}$$

Thirdly, J. Russell discovered that the decay of one large wave into several waves is possible. Fourthly, he noted that in experiments only waves of elevation are observed. He also drew attention to the fact that the solitary waves open to him pass through each other without any changes, just like the small waves formed on the surface of the water.

After 10 years, in 1844, J. Russell published his research in a paper entitled "Report on Waves" [Russell, 1844]. The work caused a cautious reaction among scientists, but it was noticed by G. Airy and J. Stokes. G. Airy criticized the results of the experiments. He noted that from the theory of long waves in shallow water, J. Russell's conclusions are not obtained, and argued that long waves cannot remain unchanged. J. Stokes also disagreed with the results of observations of J. Russell, and critically reacted to the fact of the existence of a solitary wave.

After a negative attitude to the discovery of the solitary wave for a long time it has not been recalled. The contradiction between the experiments of J. Russell and the theory of shallow water of G. Airy (and also the Stokes expansions for oscillating waves of constant form) was solved by J. Boussinesq [Boussinesq, 1871] and J. U. Rayleigh [Rayleigh, 1876]. They independently found an analytical formula for elevating the free surface on the water in the form of a square of hyperbolic secant and calculated the velocity of propagation of a solitary wave on water. Later experiments of J. Russell were repeated by other researchers, and received confirmation.

J. Boussinesq and J. W. Rayleigh showed that the correct allowance for the acceleration in the vertical direction, which determines the dispersion and which is not taken into account in the theory of shallow water of G. Airy, and also the finite amplitude, leads to a solution

$$\zeta = a\,\mathrm{sech}^2\left(\frac{x-ct}{\lambda}\right), \varepsilon = \frac{a}{H} << 1, \gamma = \left(\frac{H}{\lambda}\right)^2, \tag{2.2.13}$$

where ε – non-linear parameter, γ – dispersion parameter.

The J. Boussinesq equations, which in their traditional form represent the displacement equations on the free surface and the averaged velocity in the horizontal direction. They are not limited to one direction of propagation, and are equivalent to equations for the free surface displacement and velocity in the horizontal direction on the flat bottom of the channel. These equations were first introduced by J. Boussinesq in the report of the French Academy in 1872 [Yuen, 1987].

The final clarity in the problem of a solitary wave came after the work of Dutch scientists D.D. Korteweg and his graduate student G. de Vries, who tried to understand the essence of the observations of J. Russell. Summarizing the method of J. Rayleigh, these scientists in 1895 derived an equation for describing long waves on water [Korteweg, de Vries, 1895].

D. Korteweg and G. de Vries hypothesized that the amplitude of the waves under consideration was much less than the depth of the basin, and the wavelength was much greater than the depth of the basin. The equation that they obtained is of the form:

$$\frac{\partial \zeta}{\partial t}+c\left(1+\frac{3}{2}\frac{\zeta}{H}\right)\frac{\partial \zeta}{\partial x}+\frac{1}{6}H^2 c\frac{\partial^3 \zeta}{\partial x^3}=0\,. \qquad (2.2.14)$$

The Korteweg-de Vries equation can be obtained from the Boussineskq system of equations, a solution describing waves propagating only to the right.

The Korteweg-de Vries equation has a wave solution, known from the end of the century before last. It is expressed through a special elliptical function, studied by Karl Jacobi, which now bears his name:

$$u(x,t)=\varepsilon \times cn^2\left[\sqrt{\frac{3\gamma}{4H^2}}\times(x-Ut)\right]. \qquad (2.2.15)$$

Because of Jacobi's elliptic function, Korteweg and de Vries called these solutions *cnoidal* waves. Under certain conditions the Jacobi elliptic function becomes a hyperbolic secant and the solution of the equation is the limiting case of an infinitely large period of the wave. It is this limiting case that is a solitary wave, corresponding to the observation of J. Russell in 1834.

The solution of the Korteweg-de Vries equation is valid for describing waves only at a certain distance from the place of wave formation. At very great distances, nonlinear waves will no longer be described by the Korteweg-de Vries equation, and a more accurate model will be required to describe the process. Therefore, the Korteweg-de Vries equation is considered as an approximation (mathematical model) that corresponds, with a certain degree of accuracy, to the actual process of wave propagation on water.

For a long time, the work of Korteweg-de Vries was not noticeable. Because of its specificity, this discovery was considered quite a private fact. At that time, the physical world seemed linear and the superposition principle was considered one of the fundamental principles of most physical theories. Therefore, none of the researchers did not attach serious importance to the discovery of a solitary wave on the water.

This attitude was due to the fact that one of the fundamental physical theories – quantum mechanics, is also a purely linear theory [Landsberg, 1980]. Linearization made it possible to achieve such significant technical simplifications that it was practically considered necessary for obtaining analytical results. Nonlinear theories were perceived as incommensurably more complex and did not justify the required effort. It was assumed that it is easier to control the experimental conditions so as to always remain in linear mode.

Waves on water can be analyzed using well-known equations of hydrodynamics, which are nonlinear. Therefore, *waves in water* are also *nonlinear* in the general case. Only in the limiting case of small amplitudes, these waves can be considered linear.

Return to the discovery of a solitary wave on the water occurred to some extent accidentally. The author of this event was the famous nuclear physicist Enrico Fermi. In 1952, E. Fermi asked two young physicists S. Ulam and D. Pasta to solve one of the nonlinear problems on the first computer. They had to calculate the oscillations of 64 pendulums. In creating the initial oscillation, the researchers wanted to see how this initial fashion would be distributed across all other modes. After carrying out calculations of this task, they did not receive the expected result on the computer, but found that pumping energy into two or three modes at the initial stage of the calculation does occur, but then there is a return to the initial state. This paradox, associated with the return of the initial oscillation, became known to several mathematicians and physicists. In particular, American physicists M. Kruskal and N. Zabusky [Zabusky, Kruskal, 1965] learned about this problem, and decided to continue the computational experiments with the model proposed by E. Fermi.

After calculating and searching for analogies, these scientists established that the equation used by E. Fermi, D. Pasta, and S. Ulam, with decreasing distance between pendulums with unlimited growth of their number, goes over into the Korteweg-de Vries equation. That is, in essence, the problem proposed by E. Fermi was reduced to a numerical solution of the Korteveg-de Vries equation. Around the same years, it was shown that the Korteweg-de Vries equation is also used to describe ion-sound waves in a plasma. Then it became clear that this equation occurs in many areas of physics and, consequently, a solitary wave, which is described by this equation, is a widespread phenomenon.

Solitary waves behave like particles, so M. Kruskal and N. Zabusky such solitary waves were called *solitons*. This special name for solitary waves, consonant with the electron, proton and many other *elementary particles*, is now generally accepted.

The discoveries related to the Korteweg-de Vries equation did not result in the discovery of a soliton. The next important step related to this remarkable equation was the creation of a new method for solving nonlinear partial differential equations. It is well known that it is very difficult to find solutions of nonlinear equations.

Until the 1960s it was believed that such equations can have only certain particular solutions satisfying specially specified initial conditions. However, the Korteweg-de Vries equation even in this case was in an exceptional position.

In 1967, American physicists C. Gardner, J. Green, M. Kruskal and R. Miura [Gardner, Kruskal, 1967] showed that the solution of the Korteweg-de Vries equation can in principle be obtained for all initial conditions. They used the transformation of the Korteweg-de Vries equation to a system of two equations, now called the Lax pair [Lax, 1968] (named after the American mathematician Peter Lax, who made a great contribution to the theory of solitons). They discovered a new method for solving a number of very important nonlinear partial differential equations. This method is called the *method of the inverse scattering problem*, since it essentially uses the solution of the quantum mechanics problem of reconstructing the potential from scattering data.

Turning to waves in deep water, as is known, they are usually distributed in groups. The cause of wave grouping was found by T. Benjamin and J. Feir [Benjamin, Feir, 1967] only in 1967. By theoretical calculations, they showed that a simple periodic wave on deep water is unstable (the Benjamin-Feir instability), and therefore the waves on the water are divided into groups due to instability. The equation, which describes the propagation of groups of waves on water, was obtained by V. E. Zakharov [Zakharov, 1968] in 1968. By that time, this equation was already known in physics and was called the nonlinear Schrödinger equation.

In 1971, V. E. Zakharov and A. B. Shabat [Zakharov, Shabat, 1971] showed that this nonlinear Schrödinger equation has solutions also in the form of solitons and can be integrated by the inverse scattering method. The solitons of the nonlinear Schrödinger equation differ from the Korteweg-de Vries solitons in that they correspond to the shape of the envelope of a group of waves.

Usually under the soliton envelope there are from 14 to 20 waves, with the average wave being the largest. This is connected with the well-known fact that the highest wave in the group on the water is between the seventh and tenth (*ninth shaft*). If a larger number of waves form in the wave group, then it will decay into several groups.

The nonlinear Schrödinger equation, like the Korteweg-de Vries equation, also has wide prevalence in the description of waves in various fields of physics. This cubic equation was proposed in 1926 by the eminent Austrian physicist E. Schrödinger to analyze the fundamental properties of quantum systems [Kudryashov, 1997]. It was originally used to describe the interaction of intra-atomic particles

$$i\frac{\partial u}{\partial t}+\frac{\partial^2 u}{\partial x^2}+\nu\,|u|^2\,u=0. \tag{2.2.16}$$

The generalized or nonlinear Schrödinger equation describes a set of phenomena in the physics of wave processes: the effect of self-focusing of a laser beam, the propagation of nonlinear waves in a plasma.

Equally popular is the topological soliton, which has an entertaining history and a vast field of applications. This soliton appears in all processes, which are described by a nonlinear equation of the form

$$\frac{\partial^2 u}{\partial x \partial t}=\sin u. \tag{2.2.17}$$

It first appeared in the nineteenth century in the geometry of N. Lobachevsky in describing surfaces of constant negative curvature and is now called the sin-Gordon equation. In 1880 the Scandinavian mathematician A. V. Bäcklund showed that this equation has special transformations (Bäcklund transforms) that allow one to find successively the analytical solutions of this equation.

In cases where nonlinear surface waves in weakly dispersive shallow water are not strictly one-dimensional, the Korteweg-de Vries equation becomes unacceptable. In this case, we need another approximate model equation, this is the Kadomtsev-Petviashvili equation

$$\frac{\partial}{\partial x}\left(\frac{\partial u}{\partial t}+\frac{3c}{2H}u\frac{\partial u}{\partial x}+\frac{1}{6}cH^2\frac{\partial^3 u}{\partial x^3}\right)+\frac{1}{2}c\frac{\partial^2 u}{\partial y^2}=0. \tag{2.2.18}$$

According to Kadomtsev and Petviashvili [Kadomtsev and Petviashvili, 1970], if the dependence on the y coordinate is weak, the Korteweg-de

Vries equation can be refined by adding a small term. They obtained this additional linear term from consideration of the dispersion relation for two-dimensional long waves.

However, by that time, the number of scientists who had not been intimidated by working with waves of large amplitude had increased significantly. In particular, the publication of the works of the Royal Society of Britain, edited by M. Lighthill [Lighthill, 1967], published new materials on nonlinear waves. Along with other original results, a simple calculation of nonlinear stability. M. Lighthill writes out the already known averaged Lagrangian for waves of small amplitude in deep water, then the Lagrangian for sharpened waves of maximum amplitude, found from the calculations of J. Michell.

M. Lighthill found the critical value of the amplitude $a/\lambda = 0{,}054$, which is 76% of the maximum value above which the Benjamin-Feir instability is not observed. Thus, according to M. Lighthill, for the Benjamin-Feir instability the amplitude of the wave must be sufficiently small, and the depth is sufficiently large.

In 1965, J. Whitham [Whitham, 1965], a former student of M. Lighthill, formulated a theory of nonlinear waves of arbitrary amplitude that were locally stationary waves. For deriving the equations of motion describing the evolution of the amplitude and phase of a slowly varying wave train, J. Whitham used the variational principle of the phase-average Lagrangian. Some special cases of application of the theory of J. Whitham to the problem of the evolution of waves in deep water were those of M. Lighthill [Lighthill, 1967], John Whitham [Whitham, 1967], Mr. Yuen and B. Lake [Yuen, Lake, 1975]. Also the nonlinear Schrödinger equation G. Yuen and B. Lake were derived using the Whitham theory.

In the late 1970s and early 1980s, rapid progress was made in numerical and experimental studies on the nonlinear dynamics of high-amplitude waves. Numerous experiments were carried out for gravity waves of large amplitude on the surface of deep water. The studies gave new information about their shape, their stability, and also about the process of wave overturning; the results of M. Lighthill on the limiting amplitude were confirmed.

It should be noted the work of M. Longuet-Higgins [Longuet-Higgins, 1966, 1978], E. Cokelet [Cokelet, 1977], which mainly relate to the one-dimensional gravity waves on the surface. In the works of M. Longuet-Higgins, using the methods of the theory of functions of complex variables, the overturning of such waves having a symmetry was studied.

Rapid progress in the studies of the nonlinear dynamics of one periodic wave has also broadened the notion of the properties of a nonlinear wave field. Since the principle of linear superposition does not hold for nonlinear wave fields, the transition from periodic waves is highly nontrivial. The first systematic study in this field performed K. Hasselmann [Hasselmann, 1962, 1968], which is considered a nonlinear energy transfer along the spectrum of waves in homogeneous ocean. He found that weak nonlinear energy transfer occurs on time scales.

Attempts to find a numerical solution of nonlinear equations raise the question of the adequacy of the discrete representation of the continuous. It has long been observed that sometimes the results of the study of discrete systems do not meet the continuum limit (F. Bretherton [Bretherton, 1964]). Some preliminary results of E. Caponi *et al.* [Caponi, 1982], relevant to this issue, confirm this fact. In particular, E. Kaponi *et al.* numerically investigated the one-dimensional discrete Zakharov equation. They found that when more than one unstable mode is taken into account in the calculations, a "localized chaos" phenomenon appears in the solution in the sense that the development of these unstable modes over time becomes extremely irregular.

This means that the solution can be very sensitive to the number of modes considered in the calculation, even if the initial conditions remain unchanged in different cases. Therefore, the representation of a wave field by a finite number of modes becomes questionable when it claims information about a continuous system.

From modern books on the description of nonlinear waves, one can note the monograph of E. Infeld and J. Rowlands [Infeld, Rowlands, 2006]. The work considers linear effects, linear dispersion of waves, typical instabilities. Next, the influence of nonlinearity is considered and key nonlinear models are derived: the Korteweg-de Vries equation, its two-dimensional generalization in the form of the Kadomtsev-Petviashvili equation, the nonlinear Schrödinger equation. In the main, solutions are constructed for these equations in the form of nonlinear waves and solitons.

2.3 Investigation of Surface Gravity Waves

Basic Equations. To describe surface gravity waves, we will use the technique given in the book by L. M. Brekhovskikh and V. V. Goncharov [Brekhovskikh, 1982]. This is one of the rare books where rigorous mathematics is used to derive equations. When gravity waves are perturbed by a plane free surface of a liquid, the force of gravity is the returning force. In the boundary condition (1.6.7) its influence is described by a term $g\rho_0(0)\zeta$, equal to the pressure of a liquid column of height ζ. We eliminate in our equations the action of surface tension forces, Coriolis forces, Archimedean forces, putting $\sigma = 0,\ \ \Omega = 0,\ \ \rho_0 = const\ \ (N^2 = 0, \rho = 0)$ accordingly for this, and we introduce the notation: $\mathbf{u} \equiv \{u, \vartheta\}$ – the horizontal component of the particle velocity,

$$\nabla_{\_} = \mathbf{e}_x \partial/\partial x + \mathbf{e}_y \partial/\partial y\,.$$

Then from (1.6.11) we obtain equations for gravity surface waves:

$$\frac{\partial \mathbf{u}}{\partial t} + \frac{\nabla_{\_} p}{\rho_0} = 0, \quad \frac{\partial w}{\partial t} + \frac{1}{\rho_0}\frac{\partial p}{\partial z} = 0, \quad \nabla_{\_}\mathbf{u} + \frac{\partial w}{\partial z} = 0 \tag{2.3.1}$$

From these equations, the horizontal velocity **u** and pressure *p*, can be eliminated by expressing the latter through the vertical component of the particle velocity *w*.

$$\frac{\partial \nabla_{\_}\mathbf{u}}{\partial t} = -\nabla_{\_}\frac{\partial^2 w}{\partial t \partial z}, \quad \nabla_{\_} p = \rho_0 \frac{\partial^2 w}{\partial t \partial z} \tag{2.3.2}$$

and to obtain equation

$$\frac{\partial}{\partial t}\Delta w = 0\,. \tag{2.3.3}$$

The boundary condition on the free surface $z = 0$ for w follows easily from (1.6.7), where $\sigma = 0$ and p is taken from (2.3.2):

$$\left(\frac{\partial^3 w}{\partial t^2 \partial z} - g\Delta_{\_} w\right)_{z=0} = 0 \tag{2.3.4}$$

In addition, assuming, at the level $z = -H$ the presence of a boundary of a liquid with an absolutely solid bottom (Figure 2.3.1), we also add the condition that the normal (vertical) component of the particle velocity of the liquid vanish at $z = -H$:

$$w\big|_{z=-H} = 0 \tag{2.3.5}$$

Harmonic waves. We seek a solution of (2.3.3) satisfying the boundary conditions (2.3.4) and (2.3.5) in the form of a harmonic wave propagating in the horizontal direction:

$$w(x, y, z, t) = \Phi(z)\exp[i(\mathbf{kr} - \omega t)] \tag{2.3.6}$$

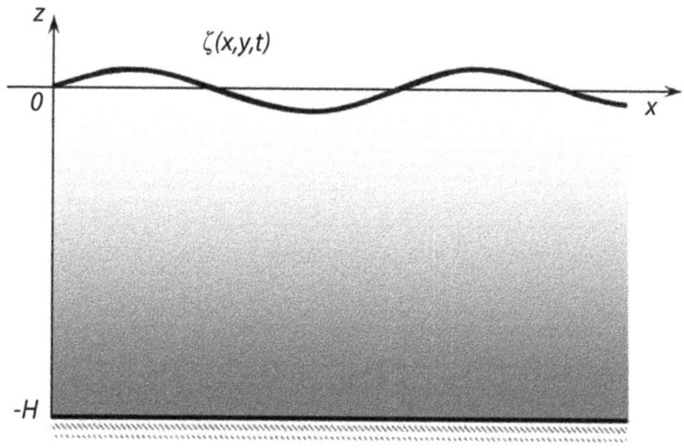

Figure 2.3.1 Geometry of boundaries.

where $\mathbf{k}=\{k_x,k_y\}$ – wave vector; $\mathbf{r}=\{x,y\}$. Substituting this expression in (2.3.3) – (2.3.5), we obtain the boundary value problem for the function $\Phi(z)$:

$$\frac{d^2\Phi}{dz^2}-k^2\Phi=0,\quad \left(\frac{d\Phi}{dz}-\frac{gk^2}{\omega^2}\Phi\right)_{z=0}=\Phi(-H)=0$$

The boundary condition $z=-H$ is satisfied if we choose the solution of the equation for Φ in the form

$$\Phi(z)=b\frac{\operatorname{sh}k(z+H)}{\operatorname{sh}kH} \tag{2.3.7}$$

Substitution of the latter in the boundary condition for $z=0$ shows that the wave (2.3.6) does not exist for arbitrary values k and ω, but only for those satisfying the dispersion relation

$$\omega^2=gk\operatorname{th}kH \tag{2.3.8}$$

It follows that the phase velocity of the gravity surface wave depends on the frequency (wave number):

$$c_{ph}=\omega/k=\sqrt{gH}\times\sqrt{\operatorname{th}kH/kH} \tag{2.3.9}$$

Consequently, the gravity surface wave propagates with dispersion, which, in particular, leads to a change in the shape of the nonharmonic surface wave. Then the group velocity of the gravity surface wave, that is, the propagation velocity of the envelope of the spectrally narrow wave packet, will be determined by expression:

$$c_{gr} = \frac{d\omega}{d\kappa} = \frac{c_{ph}}{2}\left(1 + \frac{kH}{\text{th}\, kH} - kH\,\text{th}\, kH\right). \tag{2.3.10}$$

It follows that the group velocity also depends on the wave number k. The displacement of the surface $\zeta(x, y, t)$ is determined from the boundary condition (1.6.7)

$$\frac{\partial \zeta}{\partial t} = -i\omega\zeta = w\big|_{z=0} = b\exp[i(\mathbf{kr} - \omega t)],$$

whence

$$\zeta = a\exp[i(\mathbf{kr} - \omega t)], \quad a = ib/\omega \tag{2.3.11}$$

Pulsations of pressure at arbitrary depth z, caused by the wave motion, can be found from (2.3.2), whence, with allowance $\Delta_{\_} = -k^2$ we obtain

$$p = \rho_0 \frac{i\omega}{k\,\text{th}\, kH} b \frac{\text{ch}\, k(z+H)}{\text{ch}\, kH} \exp[i(\mathbf{kr} - \omega t)]. \tag{2.3.12}$$

If we use the dispersion relation (2.3.8), then the last expression can be written in this form:

$$p = \rho_0 g \frac{ib}{\omega} \frac{\text{ch}\, k(z+H)}{\text{ch}\, kH} \exp[i(\mathbf{kr} - \omega t)] =$$

$$= \rho_o g a \frac{\text{ch}\, k(z+H)}{\text{ch}\, kH} \exp[i(\mathbf{kr} - \omega t)].$$

Further, from equations (2.3.2) it is easy to find the horizontal velocity of fluid particles in the wave

$$\mathbf{u} = i\frac{\mathbf{k}}{k} b \frac{\text{ch}\, k(z+H)}{\text{ch}\, kH} \exp[i(\mathbf{kr} - \omega t)], \tag{2.3.13}$$

direction of which coincides with the direction of propagation of the wave **k**.

If denoted by ξ the horizontal displacement of the liquid particles, it also coincides with the direction k,

$$(d\xi / dt = -i\omega\xi = \mathbf{u})$$

and accordingly through η– the vertical displacement of the particles $(d\eta / dt = -i\omega\eta = w)$, then, putting $b = \omega A \exp[i(\alpha - \pi / 2)]$ $(a = A\exp(i\alpha))$, A– is the displacement amplitude of the free surface) and selecting the real parts, we obtain:

$$\xi = -A\frac{\mathbf{k}}{k}\frac{\operatorname{ch} k(z+H)}{\operatorname{sh} kH}\sin(\mathbf{kr} - \omega t + \alpha),$$

$$\eta = A\frac{\operatorname{sh} k(z+H)}{\operatorname{sh} kH}\cos(\mathbf{kr} - \omega t + \alpha)$$

Eliminating now the dependence on t, we find the trajectories of the motion of liquid particles in a gravity surface wave:

$$\frac{\xi^2}{a_\xi^2} + \frac{\eta^2}{a_\eta^2} = 1, \quad a_\xi = A\frac{\operatorname{ch} k(z+H)}{\operatorname{sh} kH}, \quad a_\eta = A\frac{\operatorname{sh} k(z+H)}{\operatorname{sh} kH}, \tag{2.3.14}$$

are *ellipses* with the ratio of semiaxes $a_\eta / a_\xi = \operatorname{th} k(z+H)$.

In the case of a standing surface wave formed by two traveling waves of the type (2.3.6) with the same frequencies ω and amplitudes b, but with oppositely directed wave vectors **k**

$$w = 2b\frac{\operatorname{ch} k(z+H)}{\operatorname{sh} kH}\cos(\mathbf{kr})\exp(-i\omega t)$$

a similar calculation of particle mixing leads to expressions:

$$\xi = 2A\frac{\mathbf{k}}{k}\frac{\operatorname{ch} k(z+H)}{\operatorname{sh} kH}\sin(\mathbf{kr})\cos(\omega - \alpha t),$$

$$\eta = -2A\frac{\operatorname{sh} k(z+H)}{\operatorname{sh} kH}\cos(\mathbf{kr})\cos(\omega - \alpha t)$$

The trajectories of the particles are *straight lines*

$$\eta = -\operatorname{th} k(z+H)\operatorname{ctg}(\mathbf{kr})\xi, \qquad (2.3.15)$$

the slope of which depends both on the vertical coordinate *z*, and on the horizontal *x,y* coordinates. From this it is clear that for $\mathbf{kr} = 0, \pm\pi; \pm 2\pi$,... the particles move along vertical straight lines, and at $\mathbf{kr} = \pm\pi/2, 3\pi/2\pm$, ... - on horizontal.

On the basis of the expressions for the velocities of the particles of the liquid (2.3.6), taking into account (2.3.7) and (2.3.13), it is easy to verify that rot $\mathbf{v} = \mathbf{0}$, hose the motion in the surface wave is potential: $\mathbf{v} = \nabla\phi$, where

$$\phi = \frac{b}{k}\frac{\operatorname{ch} k(z+H)}{\operatorname{sh} kH}\exp[i(\mathbf{kr}-\omega t)]. \qquad (2.3.16)$$

Approximations of shallow and deep water. Usually in the theory of gravity surface waves two extreme cases are considered, differing by the relationship between *H* and the wavelength $\lambda = 2\pi/k$.

1. *Shallow water* — $H/\lambda << 1$ $(kH << 1)$. In this case, in (2.3.8) and (2.3.9) we have, $\operatorname{th} kH \approx kH$, as a result

$$\omega = \sqrt{gH}k, \quad c_{ph} = \sqrt{gH} \qquad (2.3.17)$$

The wave dispersion is missing (c_{ph} = const). For the velocity components from (2.3.7) and (2.3.13), in the same approximation we have:

$$w = b\left(1+\frac{z}{H}\right)\exp[i(\mathbf{kr}-\omega t)], \quad \mathbf{u} = i\frac{\mathbf{k}}{k}b\frac{\exp[i(\mathbf{kr}-\omega t)]}{kH} \qquad (2.3.18)$$

From this, in particular $|\mathbf{u}| >> |w|$, it follows that, that is, the velocity is practically constant over the cross section of the liquid layer and is directed parallel to the bottom. This fact is also easily established by analyzing the trajectories (2.3.14).

The ratio of the semiaxes of the ellipse $a_\eta / a_\xi = k(z+H) < kH << 1$, so that the liquid particles in the surface wave on shallow water move along elliptical orbits strongly elongated along the horizontal axis.

2. *Deep water* – $H / \lambda >> 1$ $(kH >> 1)$. In this case, we should put in (2.3.8), which leads th $kH \approx 1$, to the *dispersion relation* for waves in *deep water*:

$$\omega^2 = gk \tag{2.3.19}$$

For the phase and group velocities, we find:

$$c_{ph} = \sqrt{\frac{g}{k}} = \frac{g}{\omega}, \qquad c_{gr} = \frac{d\omega}{dk} = \frac{1}{2}\sqrt{\frac{g}{k}} = \frac{c_D}{2} \tag{2.3.20}$$

From the expressions (2.3.6), (2.3.7), (2.3.12) and (2.3.13), neglecting everywhere the quantities exp(–*kH*) in comparison with unity, we obtain:

$$\begin{aligned} w &= b\exp(kz)\exp[i(\mathbf{kr} - \omega t)], \\ \mathbf{u} &= i\frac{\mathbf{k}}{k}b\exp(kz)\exp[i(\mathbf{kr} - \omega t)], \\ p &= \rho_0\frac{i\omega}{k}b\exp(kz)\exp[i(\mathbf{kr} - \omega t)]. \end{aligned} \tag{2.3.21}$$

Thus, all perturbations of the liquid in the surface wave in deep water decrease exponentially upon deepening. The wave motion is practically missing already at a depth of the order of the wavelength λ. For the motion trajectories of the fluid particles (2.3.14) we have $a_\xi = a_\eta = A\exp(kz)$, which gives $\xi^2 + \eta^2 = A^2\exp(2kz)$ – circles whose radius decreases exponentially with depth.

Any given layer of fluid depth *H* depending on the length of propagating waves can be either shallow or deep water. Very long waves always propagate without dispersion. Examples of such waves are tidal waves or tsunami waves in the ocean. It is convenient to depict the dispersion relation (2.3.8) graphically (Figure 2.3.2), where the linear one without the dispersion part in the vicinity of the origin (long waves) is clearly noticeable, and then the dispersion corresponding to $kH >> 1$ waves in deep water.

Energy of waves. The magnitude of the energy of a wave of any kind is usually defined as the excess of the energy of a certain volume of a liquid perturbed by a wave above its value motionless. In the case of surface waves, it is convenient to choose the liquid column bounded in the vertical direction $z = -H$ and the free surface in the direction of motion of the wave, which we assume to be coincident with the axis *x*, – two planes

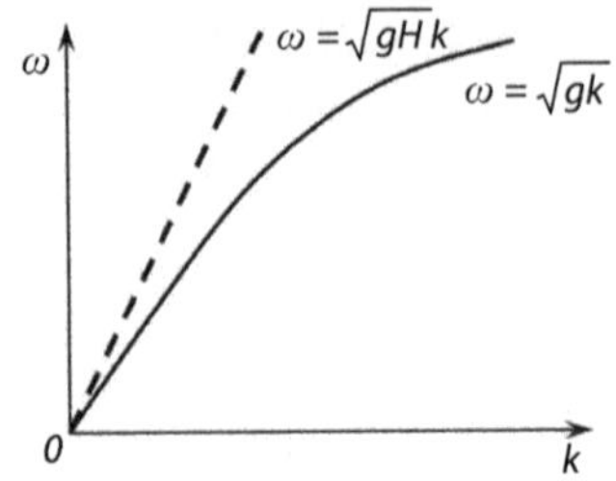

Figure 2.3.2 Dispersion ratio for shallow and deep water.

$x = x_0$, $x = x_0 + \lambda$ (λ–wavelength), in the transverse direction, analogous planes are spaced apart by one unit length.

In this case, for the kinetic energy of our volume, we have

$$E_k = \frac{\rho_0}{2}\int_{x_0}^{x_0+\lambda}\int_{-H}^{0} \vartheta^2 dxdz$$

The square of the velocity of the fluid particles, taking into account the expressions (2.3.6), (2.3.7) and (2.3.13), is written in the form

$$\vartheta^2 = B^2 \frac{\text{sh}^2 k(z+H)}{\text{sh}^2 kH}\cos^2(kx - \omega t + \alpha') + B^2 \frac{\text{ch}^2 k(z+H)}{\text{sh}^2 kH}\cos^2\left(kx - \omega t + \alpha' + \frac{\pi}{2}\right),$$

where $B\exp(i\alpha') = b$. Recall that when calculating a quadratic value ϑ_2 only the real part of the complex expressions. Returning to the calculation of the kinetic energy, we note that

$$\int_{x_0}^{x_0+\lambda} \cos^2\left[\frac{2\pi x}{\lambda} - f(t)\right]dx = \frac{\lambda}{2},$$

where $f(t)$ – arbitrary function. Therefore

$$E_k = \frac{\rho_0\lambda}{4}\frac{B^2}{\text{sh}^2 kH}\int_{-H}^{0}[\text{sh}^2 k(z+H) + \text{ch}^2 k(z+h)]dz =$$

$$= \frac{\rho_0\lambda}{4}\frac{B^2}{\text{sh}^2 kH}\int_{-H}^{0}\text{ch}[2k(z+H)]dz$$

and finally after integration we obtain

$$E_k = \frac{\rho_0}{4}\frac{B^2}{k\,\text{th}\,kH} = \frac{\rho_0 \lambda g}{4}\frac{B^2}{\omega^2} = \frac{\rho_0 \lambda g}{4}A^2. \qquad (2.3.22)$$

Here, the dispersion relation (2.3.8) is taken into account, and instead of $B = |b|$ using the displacement amplitude of the free surface $A = |a|$.

Similarly, for the increment of the potential energy, we have

$$E_n = \int_0^\lambda \left(\int_{-H}^{\zeta} \rho_0 gzdz - \int_{-H}^{0} \rho_0 gzdz \right) dx$$

$$= \rho_0 g \int_0^\lambda \int_0^\zeta gzdz = \frac{\rho_0 g}{2} \int_0^\lambda \zeta^2 dx$$

Substituting in place of the integral *t*, the real part of expression (2.3.11) $A\cos(kx - \omega t + \alpha)$ and integrating, we obtain

$$E_n = \frac{\rho_0 \lambda g}{4} A^2 = E_k. \qquad (2.3.23)$$

Thus, the kinetic and potential energies in the progressive surface wave are equal to each other and are constant in time.

Gravitational-Capillary Waves on the Surface of a liquid

Pure capillary waves. Consider the waves due to surface tension forces. The action of these forces is described by a term - $\sigma\Delta\_\xi$ in the boundary condition (1.6.7). Eliminating the wave motions of other types ($\rho_0 = const,\ N^2 = 0, \Omega = 0, g = 0$) in (1.6.7) and (1.6.11), we again obtain equation (2.3.3) for the vertical velocity component. However, the boundary condition on the unperturbed free surface $z = 0$ will be different. Namely, putting in the second equation (1.6.7) $g = 0$ and acting on it by the operator $\partial\Delta\_/\partial t$ we have

$$\frac{\partial}{\partial t}\Delta\_ p\Big|_{z=0} = -\sigma\Delta\_\Delta\_\frac{\partial\zeta}{\partial t}$$

Hence, expressing $\partial\zeta/\partial t$ in w terms of the first condition (1.6.7) and Δ_p the just (and, in this case, relation (2.3.2)), we find

$$\left(\frac{\partial^3 w}{\partial t^2 \partial z}+\gamma\Delta\_\Delta\_w\right)_{z=0}=0, \tag{2.3.24}$$

where $\gamma=\sigma/\rho_0$.

Capillary waves are high-frequency waves with a small wavelength. Therefore, in practical problems one can suppose $kH\to\infty$, as in the case of an infinitely deep fluid. The solutions of equation (2.3.3) must then vanish at $z\to-\infty$. If we seek a solution in the form of a harmonic wave (2.3.6), then for $\Phi(z)$ and $w(x,y,z,t)$ we have:

$$\Phi(z)=b\exp(kz), \quad w=b\exp(kz)\exp[i(\mathbf{kr}-\omega t)].$$

Substituting w in the boundary condition (2.3.24), we obtain *dispersion relation for capillary waves*

$$\omega^2=\gamma k^3 \tag{2.3.25}$$

In this case, the phase and group velocities of the waves are equal to:

$$c_{ph}=\sqrt{\gamma k}, \quad c_{gr}=\frac{3}{2}\sqrt{\gamma k}=\frac{3}{2}c_{ph}. \tag{2.3.26}$$

Expressions for the pressure p and the horizontal velocity component coincide with the corresponding expressions for gravity waves in deep water (2.3.21). Therefore, the trajectories of the liquid particles in the capillary wave are circles.

Gravitational-capillary waves. We now take into account the joint action of gravity and surface tension. In the homogeneous ($\rho_0=const$), incompressible ($A=\infty$), and nonrotating ($\Omega=0$)liquid, the equations (2.3.1) – (2.3.3), the boundary condition (2.3.5), and also the solutions (2.3.6), (2.3.7), (2.3.11) – (2.3.13) in the form of harmonic waves.

All the difference from the cases considered above consists in using the complete dynamic boundary condition for p the free surface (1.6.7), which w for the combination of conditions (2.3.26) and (2.3.47):

$$\left(\frac{\partial^3 w}{\partial t^2 \partial z}-g\Delta\_w+\gamma\Delta\_\Delta\_w\right)_{z=0}=0$$

Accordingly, only the dispersion relation for the waves will change

$$\omega^2 = (gk + \gamma k^3) th kH . \tag{2.3.27}$$

Hence we find the square of the phase velocity

$$c_{ph}^2 = (g / k + \gamma k) th kH. \tag{2.3.28}$$

In particular, for waves in deep water ($kH >> 1$) we have:

$$\omega^2 = gk + \gamma k^3, \quad c_{ph}^2 = g / k + \gamma k . \tag{2.3.29}$$

For small k (long waves), the first term on the right is the first term, and we already have the case of gravity waves (2.3.8) already studied. The phase velocity decreases with increasing k. On the contrary, for large k (short waves) in (2.3.29), only the second term, capillary waves, should be retained. The phase velocity increases with increasing k. For some $k = k_0 = 2\pi / \lambda_0$ the phase velocity of the wave will be minimal. Equating the derivative of the right-hand side of the second equation (2.3.29) to zero, we find:

$$k_0 = \sqrt{g / \gamma}, \quad \lambda_0 = 2\pi \sqrt{\gamma / g}, \quad (c_{ph})_{\min} = \sqrt{2} (g\gamma)^{1/4} . \tag{2.3.30}$$

For the water-air boundary $\gamma \approx 73 Am^3 / s^2$ we obtain $\lambda_0 = 1{,}714 \text{ cm}$, the corresponding frequency $f_0 = (c_{ph})_{\min} / \lambda_0 = 13{,}5 \text{ Hz}$ and $(c_{ph})_{\min} = 23{,}1 \text{ cm/s}$.

Note that the wavelength λ_0 is boundary, separating gravity and capillary waves. At $\lambda >> \lambda_0$ prevailing gravity forces, with $\lambda << \lambda_0$ – capillary. For the group velocity of gravitational-capillary waves in deep water ($kH >> 1$), differentiating (2.3.29) on k, we have

$$2\omega c_{gr} = g + 3\gamma k^2$$

consequently,

$$c_{gr} = \frac{g + 3\gamma k^2}{2\omega} = \frac{c_{ph}}{2} \frac{g / \gamma + 3k^2}{g / \gamma + k^2} = \frac{c_{ph}}{2} \frac{k_0^2 + 3k^2}{k_0^2 + k^2} . \tag{2.3.31}$$

From this, we obtain, as it should be, for gravity waves ()the expression (2.3.10) for $kH >> 1$, and for capillary waves $k_0 << k$ – expression (2.3.49).

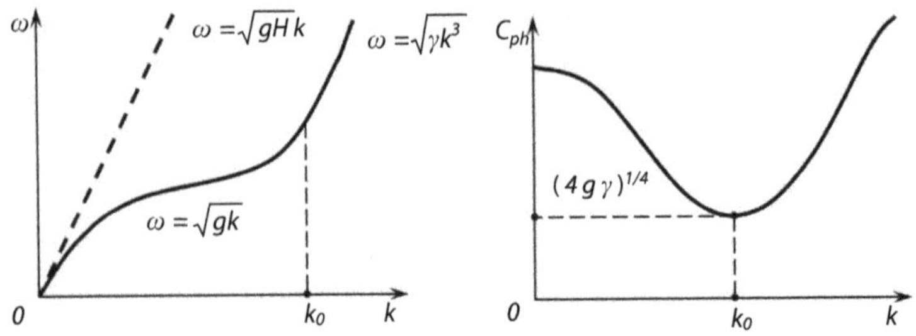

Figure 2.3.3 Dispersion dependence for gravitational-capillary waves.

Figure 2.3.3 shows a schematic depiction of dispersion dependences $\omega(k)$ и $c_\phi(k)$ for gravitational capillary waves.

2.4 Spatial modeling of Wave Phenomena on Shallow Water Surface

To study the wave processes occurring in shallow water areas, consider three-dimensional spatial models of some wave phenomena on the shallow water surface in the *linear* case.

Figures 2.4.1 and 2.4.2 show spatial models of the propagation of the surface gravity wave, as well as the superposition of these waves in shallow water. The graphs are given for a single level of the pressure amplitude on the basis of the expression (2.3.12). The superposition of two waves simulates a situation where waves exist on the surface of the sea that come from different directions or as a standing wave when reflected from a solid vertical rock. As a result of superposition on the surface, separate three-dimensional cells.

Figure 2.4.3 shows a spatial model of a packet of surface gravity waves propagating at an angle of $\theta = 45°$ to the y axis *y*.

Figure 2.4.4 shows the spatial model of the overlap of two packets arriving at angles of 45° and 90° to the x-axis. Here, we again observe individual three-dimensional cells; however, their arrangement is already different from the previous case 45° and 90° to the x-axis *x*. Some local areas appear with a reduced level of cells. Spatially three-dimensional models were created by the method of plane-parallel transfer of the initial pressure field along the y axis [Lavrent'ev, 1973].

To model the wave phenomena that occur when surface gravity waves approach the shore, or rather, the phenomenon of *refraction* – distortion

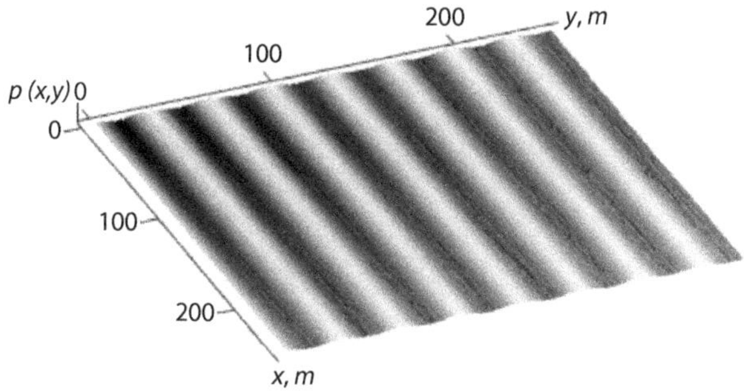

Figure 2.4.1 Spatial model of the surface gravity wave at: $f = 0.3$Hz; $\lambda = 33$ m; $H = 10$ m; $c = 9.9$ m/s.

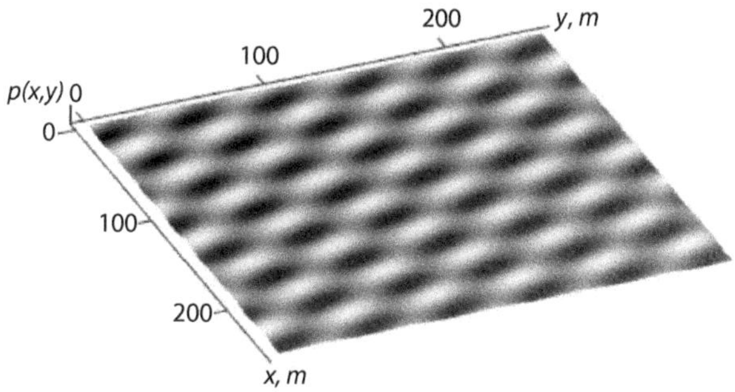

Figure 2.4.2 Spatial model of superposition of two surface gravity waves at: $f = 0.3$ Hz; $\lambda = 33$ m; $H = 10$ m; $c = 9.9$ m/s, $\theta = 0^0$ and 45^0.

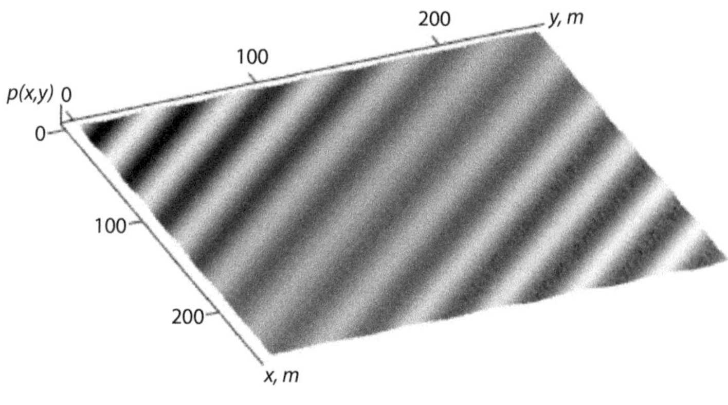

Figure 2.4.3 Spatial model of a packet of surface gravity waves at: $f = 0{,}3$; $0{,}33$ Hz; $\lambda = 33$; 30 m; $H = 10$ m; $c = 9.9$ m/s, $\theta = 45^0$.

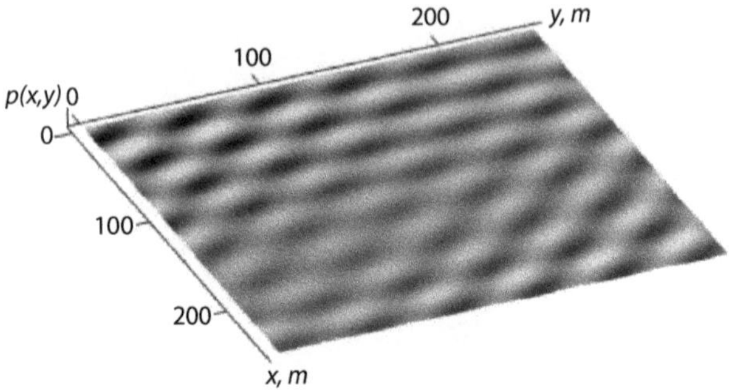

Figure 2.4.4 Spatial model of superposition of two packets of surface gravity waves at: $f = 0.3; 0.33$ Hz; $\lambda = 33; 30$ m; $H = 10$ m; $c = 9.9$ m/s, $\theta = 90^0$ and 45^0.

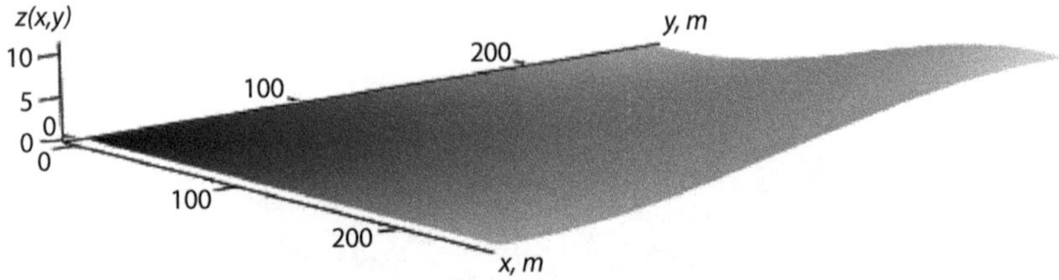

Figure 2.4.5 Spatial model of the cape (250 × 250 × 10 m) according to the analytical expression $z(x,y) = 0{,}0002xy \times \sin(0{,}00003xy)$.

of the wave front, we will create three-dimensional models of shore formations. We model with the help of analytical functions a cape, a bay and a flat coastline.

Figure 2.4.5 shows the model of a three-dimensional cape with dimensions 250 × 250 × 10 m (scale on the z increased). It is described by an analytical expression

$$z(x, y) = axy\sin(xy). \qquad (2.4.1)$$

Figure 2.4.6 shows the spatial model of a straight coastal protrusion described by an analytical expression

$$z(x, y) = ab(y)x^2. \qquad (2.4.2)$$

Figure 2.4.7 shows the spatial model of the bay according to the analytical expression

$$z(x, y) = x^2\cos(y). \qquad (2.4.3)$$

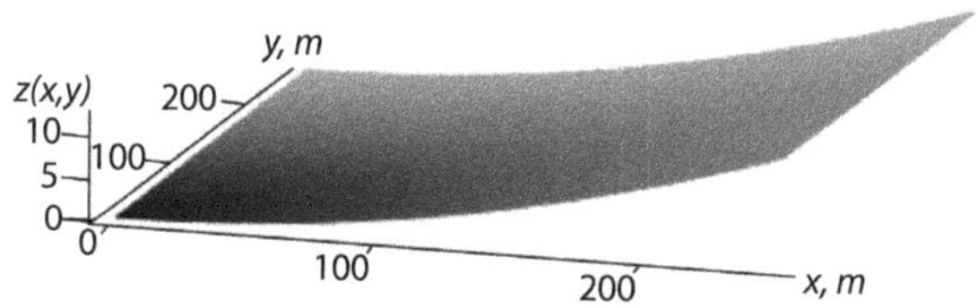

Figure 2.4.6 The spatial model of the shore protrusion (250 × 250 × 10 m) according to the analytical expression $z(x,y) = 0{,}0002yx^2$. (250 × 250 × 10 m)

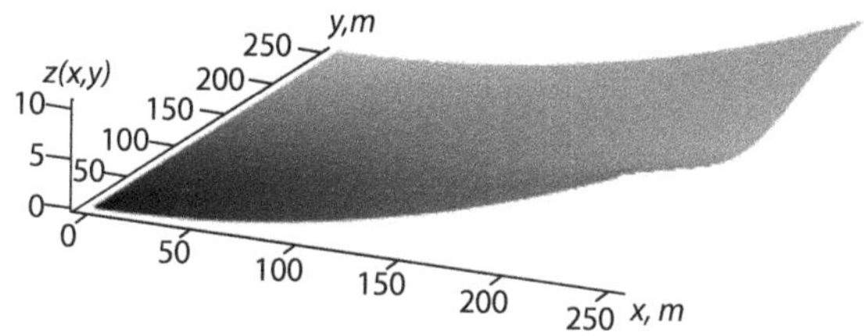

Figure 2.4.7 Spatial model of the bay (250 × 250 × 10 m) by analytical expression $z(x,y) = 0{,}00017x^2 \times (\cos(0{,}05y) + 3))$.

To simulate the process of refraction of surface gravity waves on these shore formations, let us use the expression for the wave number in the two-dimensional case (based on the expression (2.3.17)):

$$k(x,y) = \sqrt{\frac{\omega^2}{g(H - z(x,y))}}, \tag{2.4.4}$$

where $z(x,y)$ – analytical expression of the relevant shore formation. It should be noted here that the expression (2.3.17) was obtained under the condition of a constant depth, but it is also applicable to a variable depth with an insignificant slope steepness [Shuleikin, 1954]. In our case, the steepness of the slope is within 0.1^0, despite the fact that the depth varies from point to point, according to the calculated grid, it is constant at each point.

The length of the surface of the bay along the x axis is limited by the coastline, the wave source is located at the origin. Waves on the surface of the bay are propagated in the absence of wind, i.e., are swell waves. For calculations, the surface of the bay is divided into a grid with a spacing of 0,5 m along the x and y, the value of which is an order of magnitude smaller

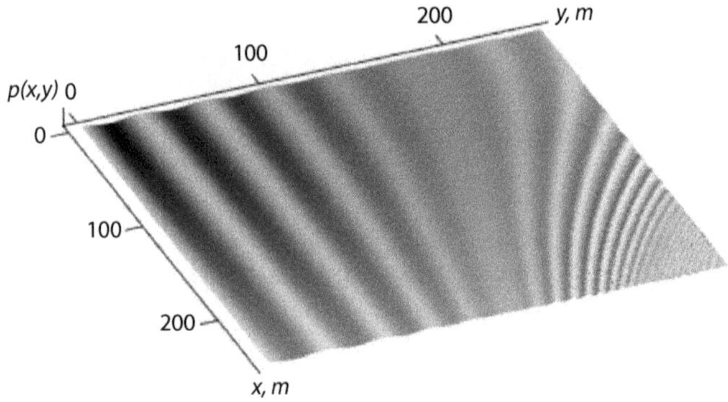

Figure 2.4.8 Spatial model of refraction of a packet of surface gravity waves on the cape at: $f = 0.3; 0.33$ Hz; $\lambda = 33; 30$ m; $H = 12$ m; $c = 11$ m/s, $\theta = 0^0$.

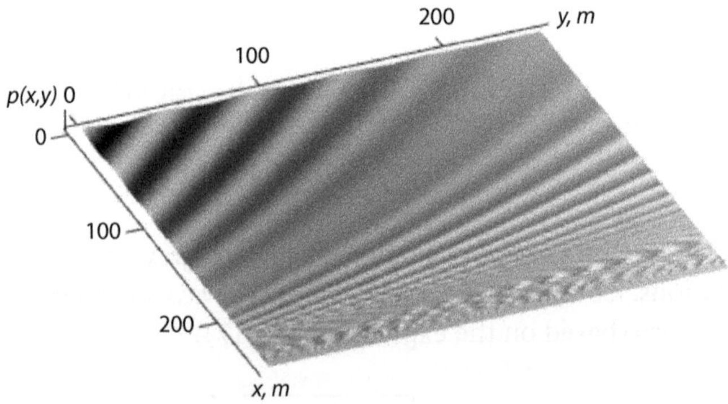

Figure 2.4.9 Spatial model of refraction of a packet of surface gravity waves on a ledge at: $f = 0.3; 0.33$ Hz; $\lambda = 33; 30$ m; $H = 12$ m; $c = 11$ m/s, $\theta = 45^0$.

than the sampling interval for these waves according to Kotel'nikov's theorem [Gonorovskii, 1957].

Figures 2.4.8, 2.4.9 and 2.4.10 show spatial models of the phenomenon of refraction of a packet of surface gravity waves on the cape with the direction of arrival $\theta = 0°$, on the flat coastal ledge $\theta = 45°$ and in the bay $\theta = 90°$, at $f = 0.3; 0.33$ Hz; $\lambda = 33; 30$ m; $H = 12$ m; $c = 11$ m/s.

With the approach to the shore beginning at some critical depths, the surface wave front begins to distort and tends to become normal to the shoreline. Distortion of the front occurs as a result of a decrease in the propagation velocity, which leads to a decrease in the wavelength at a constant frequency.

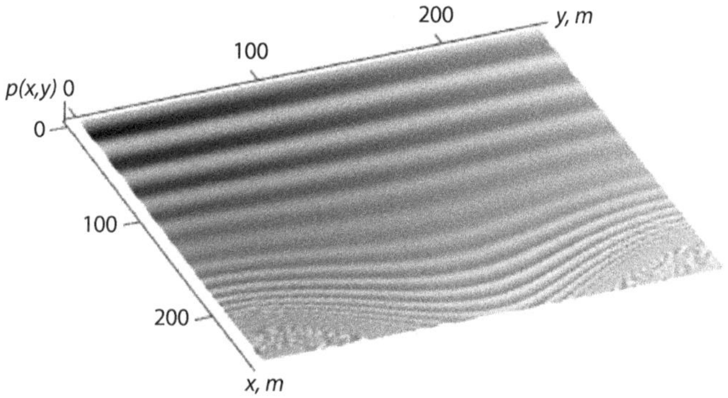

Figure 2.4.10 Spatial model of refraction of a packet of surface gravity waves in a bay at: $f = 0.3$; 0.33 Hz; $\lambda = 33$; 30 m; $H = 12$ m; $c = 11$ m/s, $\theta = 90^0$.

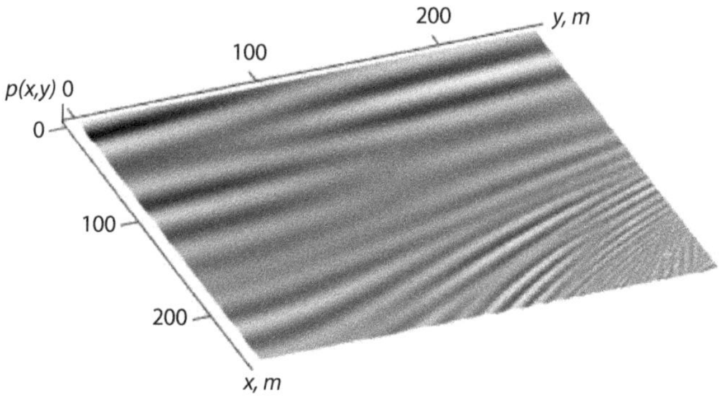

Figure 2.4.11 Spatial model of refraction of two packets of surface gravity waves on the cape at: $f = 0.3$; 0.33, (0,4; 0,44) Hz; $\lambda = 33$; 30, (25, 22) m; $H = 12$ m; $c = 11$ m/s, $\theta = 90^0$ и 85^0.

Figures 2.4.11 and 2.4.12 show spatial models of refraction to the cape of two packets of surface gravity waves with different angles of arrival at: $f = 0.3$; 0.33, (0.4; 0.44) Hz; $\lambda = 33$; 30, (25, 22) m; $H = 12$ m; $c = 11$ m/s.

Depending on the direction of arrival on the surface of the bay, three-dimensional cells of different scale and structure are formed. When overlapping packets of surface waves with angular intervals of 30^0, smaller-scale three-dimensional cellular structures are created, compared with packets of close angles of arrival.

Figures 2.4.13, 2.4.14 and 2.4.15 show spatial models of refraction at the cape, on the straight coastal protrusion and in the bay of three different surface waves at: $f = 0.3; 0.33; 0.4$ Hz; $\lambda = 33; 30; 25$ m; $H = 12$ m; $c = 11$ m/s coming from different directions. On the surface of the bay, three-dimensional

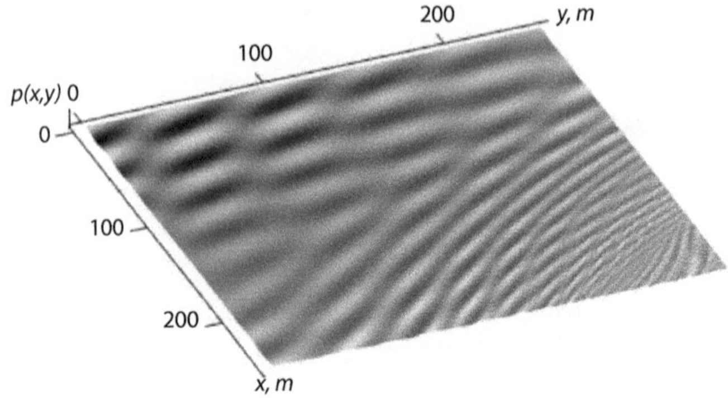

Figure 2.4.12 Spatial model of refraction of two packets of surface gravity waves on the cape at: $f = 0.3$; 0.33, (0.4; 0.44) Hz; $\lambda = 33$; 30, (25, 22) m; $H = 12$ m; $c = 11$ m/s, $\theta = 90^0$ и 60^0.

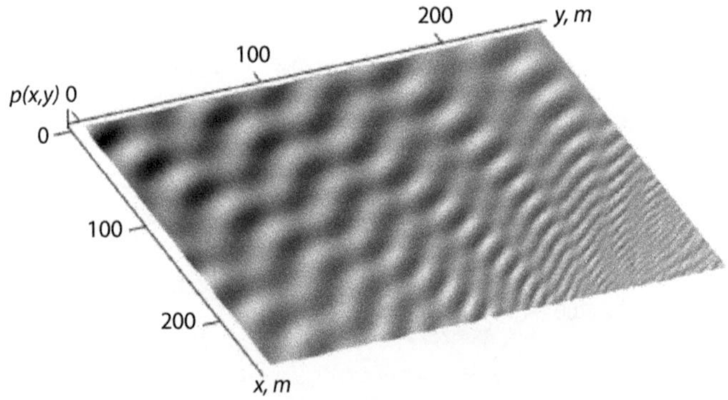

Figure 2.4.13 Spatial model of refraction of three surface gravity waves on the cape at: $f = 0.3$; 0.33; 0.4 Hz; $\lambda = 33$; 30; 25 m; $H = 12$ m; $c = 11$ m/s, $\theta = 90^0$, 60^0 and 30^0.

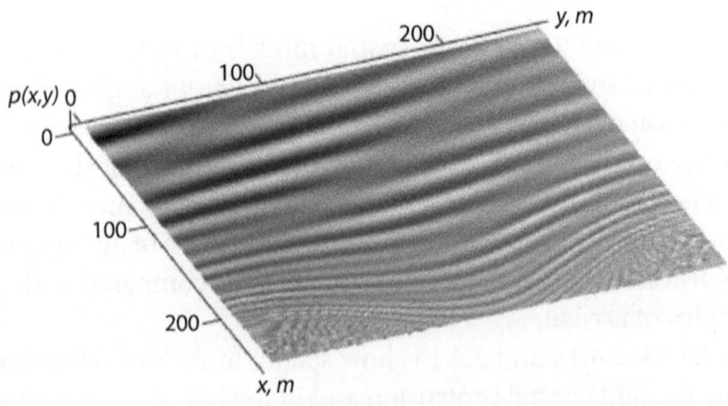

Figure 2.4.14 Spatial model of refraction of three surface gravity waves in a bay at: $f = 0.3$; 0.33; 0.4 Hz; $\lambda = 33$; 30; 25 m; $H = 12$ m; $c = 11$ m/s, $\theta = 90^0$, 85^0 and 80^0.

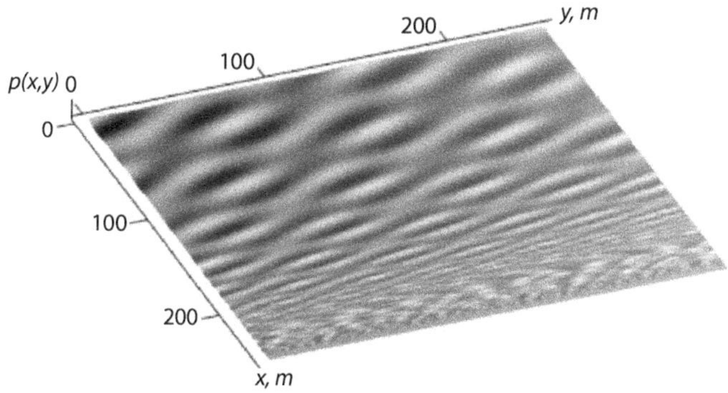

Figure 2.4.15 Spatial model of refraction of three surface gravity waves on the ledge at: $f = 0.3; 0.33; 0.4$ Hz; $\lambda = 33; 30; 25$ m; H = 12 m; $c = 11$ m/s, $\theta = 90^0, 45^0$ and 70^0.

Figure 2.5.1 The wave structure on the water surface, formed from surface waves with an infinitesimal amplitude (*Dyurso village, the Black Sea*).

wave cells having a similar structure are also formed, however, without characteristic regions with reduced levels of individual waves.

2.5 Actual Observations of Wave Phenomena on the Surface of Shallow Water

Figure 2.5.1 ÷ 2.6.5 shows the actual photographic fixation of some wave phenomena on the shallow water surface: superposition of wave, formation of periodic wave structures, refraction of surface waves in the bay, damping of surface waves on the shore, dispersal of wind waves along the shore.

Figure 2.5.2 Periodic wave cells formed by the superposition of surface waves of different lengths, small ripples on the swell (*Dyurso village*).

Figure 2.5.3 Attenuation of a surface wave with an infinitesimal amplitude on the shore (*Taganrog Bay, Azov Sea*).

2.6 Ship Waves. "Reactive" Ducks of the Alexander Garden

One of the interesting problems of the wave theory is the investigation of waves on the surface of a liquid that arise when a rigid body moves along

Figure 2.5.4 Refraction of surface waves in the bay conditions (*Dyurso village*).

Figure 2.5.5 Dispersal of wind waves along the coastline, oblique surf (*Dyurso village*).

its surface. This task is also interesting for determining the impact on the coastal structures of waves created by a moving ship. The moving ship expends energy to overcome the resistance of water, the excitation of vortices in a turbulent wake and the generation of surface waves.

A moving tail develops behind the moving ship, extending far into the region traversed by the ship, and is an overlap of two different families of waves – *transverse and longitudinal.* Similar floating traces also leave floating objects of natural origin, such as ducks in Figure 2.6.1. In this case, one can determine the angular size of the wave trail and compose the equation of fluid motion within this wave region.

The first successes in solving the problem of the formation of ship waves were achieved by Kelvin [Kelvin, 1910], [Sretensky, 1977]. In Kelvin's theory, the action of a ship on water is replaced by the action of concentrated pressure pulses applied at the points of the ship's path. The concentrated impulse moves rectilinearly at a constant rate from infinity. At each moment of time, this pulse adds known velocities to the previously formed velocities. The elevation of the liquid surface at a given point and at a given time t will thus be composed of the sum of the elevations created by all the pulses applied to the surface of the liquid for a time from ∞ up to a given moment t. Relative to the system of coordinate axes moving along with the momentum, the surface of the liquid will have an unchanged shape. The equation of this surface was first compiled and studied by Kelvin.

After Kelvin's theory ship waves were investigated in [Havelock, 1908], [Hogner, 1922]. Havelock's work contains a general description of the wave motion of the surface of a liquid behind a ship, which is replaced by a

Figure 2.6.1 Wedge-shaped wave tracks from floating ducks in the pond of the Alexander Garden (*St. Petersburg*).

moving pressure pulse. Hogner replaces the action of the ship on water, by the action of pressures moving along the surface of the liquid at a constant speed. In the works of [Petters, 1949], [Lighthill, 1957] and [Ursell, 1960] theory of ship waves has been further developed.

The most surprising result, first obtained by Kelvin, is that in deep water waves are concentrated in a wedge-shaped region with a certain angle $\theta = 19°28'$ (arcsin 1/3). This angle does not depend on the speed or shape of the object. The angle is determined only by the fact that on deep water the group velocity of the waves is equal to half the phase velocity (expression 2.3.20).

The fronts of these waves do not coincide with the lateral lines of the track, but form with it a certain angle $\alpha = 35°16'$. These waves are called *longitudinal* (Figure 2.6.2). Directly behind the stern of the ship there is another family of waves that propagate in the direction of the ship's movement with its speed. These waves are called *transverse*, since their front is perpendicular to the direction of motion of the ship. In real conditions, the

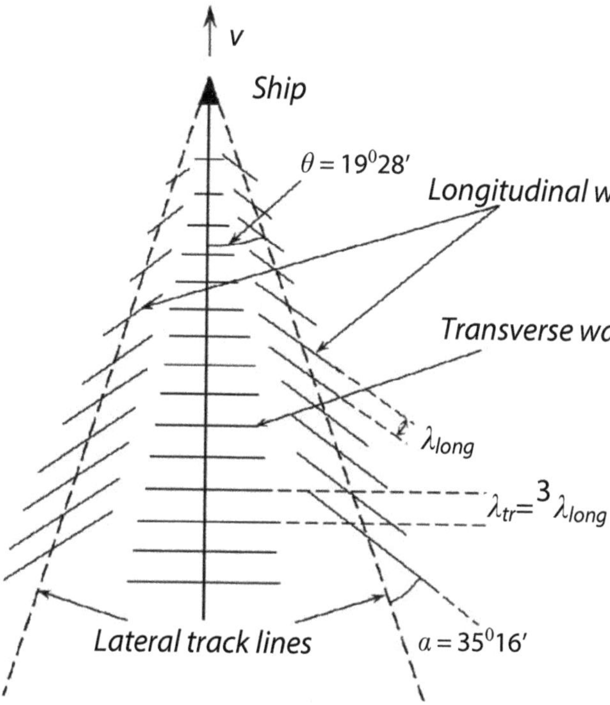

Figure 2.6.2 Ship Waves.

transverse waves are often perturbed by a turbulent wake behind a ship with a large amount of foam formed when waves crash at the bow of the ship and because of cavitation on the screws. In addition, usually wind wave waves are superimposed on the system of ship waves [Monin, Krasitsky, 1985].

Consider a ship moving with velocity v and moving from point A to point B (Figure 2.6.3) in time t [Whitham, 1976]. The crest of the wave will remain fixed, relative to the ship in the event that

$$v\cos\theta = c \tag{2.6.1}$$

where θ — angle between the normal (direction of the vector $\mathbf{k}$) and the line of motion AC. This condition is most easily understood by going into a report system in which a stream with velocity v flows around a stationary ship. In this moving reference frame, the component $v\cos\theta = c$ of the flow velocity normal to the wave element must be compensated by the phase velocity of the element under consideration. This condition allows us to determine the value of $\mathbf{k}$ in the direction θ.

Geometrically (Figure 2.6.3), this can be represented by constructing a semicircle with diameter AC and noting that $AC = vt$, $AB = vt\cos\theta = ct$. Consequently, for wave elements parallel to the chord BC, we have $ct = AB$. Further, c – is the phase velocity corresponding to the condition (2.6.1), but the location of the waves under consideration is determined precisely by the group velocity $c_{gr} = c/2$. Waves formed at point A must travel a distance $c_{gr}t = ct/2$.

Consequently, in the direction θ they will be detected at the point E – the midpoint of the segment AB. Having considered all the values θ, we get that the waves arising at the point A and contributing to the stationary picture lie on the circle of radius vt / 1 with the center at the point O, at that $OC = 3vt/4$. Finally, fixing the point C and changing t, we get the set of circles shown in Figure 2.6.4b.

Each circle has a radius equal to one-third the distance from its center to point C. Consequently, these circles fill the wedge-shaped region with the half-solution angle $\arcsin 1/3 = 19°28'$.

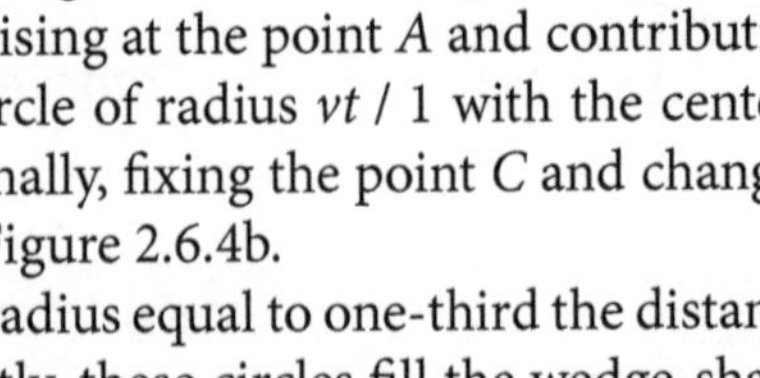

Figure 2.6.3 Construction of wave elements.

It should be emphasized that Figure 2.6.4b corresponds to a supersonic flow with a Mach number of 3, all floating objects have an effective Mach number of 3 [Whitham, 1976]. Let us assume that the medium is stationary, and the source of perturbations moves rectilinearly with velocity v. If, then $v < c$, the source moves more slowly than the perturbations produced by it, and the disturbance pattern will be as shown in Figure 2.6.4a and Figure 2.6.5 [Lavrentiev, Shabat, 1973]. If $v > c$, then the source will outstrip the disturbances, and we will get the picture shown in Figure 2.6.2b.

One of the most significant differences between supersonic and subsonic modes of the medium is related to the different nature of the propagation of local disturbances in them. Namely, at subsonic velocities, perturbations propagate throughout the space, and for supersonic velocities, only within a certain sector.

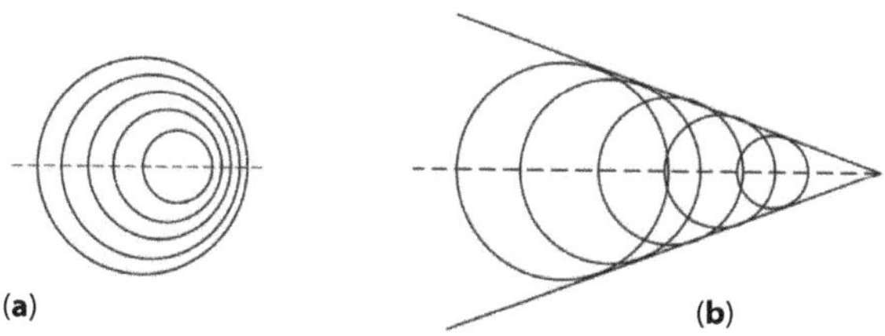

Figure 2.6.4 The envelope of perturbations arising in consecutive moments.

Figure 2.6.5 Divergent waves from the swimmer. Speed of swimmer is less than the speed of divergent waves (*Taganrog Bay*).

Conclusions

The second chapter carried out research and modeling of wave activities on the shallow water surface. The different types of waves that occur on the surface of the water and exist in shallow waters are described. A review of studies on surface gravity waves, the main historical stages of development and the peculiarities of existing water-wave theories are presented.

The main equations that describe the surface gravity waves, the main properties of the waves at shoaling, are provided. The capillary and gravitational-capillary waves are also described.

The main properties of the surface gravity waves at shallow water are examined, 3D models of wave activities at the shallow water in the linear case: distribution of the wave package, overlay the wave package with different entry directions.

In order to create a model of refraction of surface gravity waves at shoaling, the 3D models of coast formations are created: cape, bay and straight coastline. 3D models of the refraction of separate package and superposition of surface gravity waves on the various coast formations for linear case.

The waves on the surface of the fluid that emerged when the solid body was moving on its surface were considered. The main parameters of the wave tail behind the moving ship were described. The waves for cases with different ratios between the speed of a floating object and the velocity of the disconnected waves were showed.

Actual illustrations of wave activities were provided: the superposition of waves of ripples and swell, refraction of waves in the conditions of the bay, and also various cases of the distribution of ship waves.

The results of the studies in this chapter have been published in the following main works: [Abbasov, 2010 Fizmatlit], [Abbasov, 2001], [Abbasov, 2003], [Abbasov, 2004].

3

Modeling of Nonlinear Surface Gravity Waves in Shallow Water

3.1 Overview of Studies on Nonlinear Surface Gravity Waves in Shallow Water

The influence of shallow on long (tidal) waves was initially investigated by G. Airy [Lamb, 1947]. From the hydrodynamics equations, an approximation of the tidal wave profile was derived, initially a simple form of cosine and distorted by shallow water. An approximation law of increasing the second harmonic as waves spread in the shallow sea area was found. However, the Eulerian equation takes into account a nonlinear member

$$u\frac{\partial u}{\partial x},$$

which is always discarded when the small amplitude waves are studied.

In the application of the equations of motion and the continuity of the tide activity, he used *the method of successive approximations* [Airy, 1845].

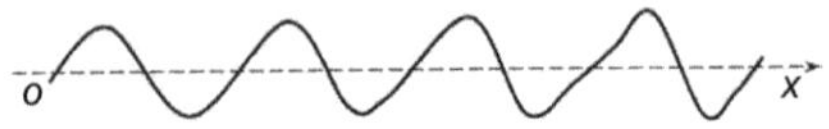

Figure 3.1.1 Tidal wave profile.

An example was the channel that one end connects to the open sea, where elevation is given by the formula

$$\eta = a\cos\omega t. \tag{3.1.1}$$

The second approximation for the elevation function is

$$\eta = a\cos\omega\left(t - \frac{x}{c_0}\right) - \frac{3}{4}\frac{g\omega a^2}{c_0^3}x\sin 2\omega\left(t - \frac{x}{c_0}\right) \tag{3.1.2}$$

Figure 3.1.1 shows (with increased amplitude) the profile of the wave calculated from the resulting solution for a certain private case. The crest and lows of the waves were found to be asymmetrical, the low was more flattering than the crest.

The presence of a multiplier x outside the sign of the trigonometric function indicates the existence of a boundary beyond which the approximation is not applicable. The condition for applying the method of successive approximations is that $g\omega a^2/c_0^3$ should be little. If you consider that $c_0 = \sqrt{gH}$, $\lambda = 2\pi c_0/\omega$, then this fraction will be equal $2\pi(a/H)(x/\lambda)$.

Hence, as little as the ratio of initial elevation to the depth, the above fraction ceases to be small, when x is sufficiently larger than the wavelength.

The solution in the second approximation consists of two members. The second is supertide, or the tide of the second order, as it is proportional a^2; this tide is twice the frequency of the initial tidal fluctuation.

If the initial disturbance at the mouth of the channel is given by the formula

$$\eta = a_1\cos\omega_1 t + a_2\cos(\omega_2 t + \phi), \tag{3.1.3}$$

in the second approximation, tidal waves with combination periods are similarly obtained. They are similar to combination tones in acoustics, which were first studied by G. von Helmholtz [Helmholtz, 1859].

As a result of two consecutive approaches, G. Airy received the shape of a profile distorted by the second harmonic. It should be noted, however, that the emergence and gradual increase in the harmonics of higher orders will inevitably reduce the amplitude of major fluctuations as a result of the transfer of more and more energy. Neglect of pumping energy from the main wave will lead to a violation of the law of energy conservation.

The method of G. Airy was used by G. Jeffreys [Jeffreys, 1924] for surface waves. He tried to find out their destroying conditions in shallow water. However, a profile of a wave spread in shallow water can be described by the sum of many higher harmonics.

Research on the surface of the water was extensively examined in the middle of the last century by L. N. Sretensky [Sretensky, 1977]. He built a theory of free tidal waves within the polar basin, created a map of the distribution of the semidiurnal tidal wave in the Earth's water hemisphere, taking into account the shape of the continents and islands of the hemisphere.

L. N. Sretensky did a lot of work on a common linear wave theory. He testified that the surface of the fluid section showed waves of greater amplitude than the free surface. He examined the unresearched task of distribution steady waves on the surface of a 3D fluid flow of infinite depth. He created an algorithm that allows you to solve, in any approximation, the dynamics task of the 3D steady waves of the finite amplitude.

He also explored wave motion on the surface of the channel with variable depth [Sretensky, 1963]. He obtained results for basins that have a downward bottom. Results obtained through analytical theory techniques and differential equations. The solution of the heterogeneous differential equation defines a wave motion that has unrestrained growth in the vicinity of the coast and is moving far from coast to standing waves.

In the work of B. N. Rumyantsev [Rumyantsev, 1960] the sustained heavy fluid motions at the inclined coast are considered. The basin floor is represented by a straight line tilting to the horizon at a certain angle. The surface of the liquid is deviating from the starting position and some initial speeds are reported to the particles of the fluid. B. N. Rumyantsev solved this task with the help of integral equations. He considered the distribution of the concentrated surface elevation of the liquid. While disintegrating, the elevation forms two groups of waves distributing in opposite directions from the center of disturbances (to the coast and from the coast). The spatial task of the waves in the sloping-bottom basin was investigated in M. Roseau's work [Roseau, 1952].

The theory of standing waves on the surface of a 3D mass of fluid was considered by Y. I. Sekerzh-Zenkovich [Sekerzh-Zenkovich, 1952]. However, the equations for the surface of the liquid are solved by an expansion method in powers of the small parameter.

A fairly extensive pilot study of wind waves in the storm basin was done by the V. V. Shuleikin [Shuleikin, 1968]. He worked on a mechanism for the early development of wind waves. With the glowing spherical floats having neutral buoyancy, he explored the orbits of particles in wind waves. He created a system of equations, graphic-analytical way to calculate the wind wave field in the deep sea. In his work [Shuleikin, 1954], he described the process of breaking the waves under the shallow and presented experimental data on the changes in the profile of wind waves at different moments. By analogy with the optical refraction, he considered in his work [Shuleikin, 1935] the refraction of the waves on the continental platform. He introduced the concept of the refraction of sea waves based on the ratio of their velocities in the infinite deep sea and at arbitrary depth.

The processes of refraction of surface waves were investigated by Yu. M. Krylov. In [Krylov, 1950] he considered the refraction of waves on the continental shelf. In its refractive problems, the main characteristic factor is the refractive index of the medium, which varies from point to point. He developed this theory with reference to the island of round and triangular shape, and to the cape to which the waves approach. This theory also makes it possible to calculate the change in the amplitude of waves approaching the shore.

At the end of the last century, a great contribution to the theory of shallow water was made by the works of D. Peregrine. For a channel with an uneven bottom, he [Peregrine, 1967] derived the Korteweg-de Vries equation:

$$\frac{\partial u}{\partial t}+u\frac{\partial u}{\partial x}+k^2\frac{\partial^3 u}{\partial x^3}=\nu(H)u, \tag{3.1.4}$$

where the function – $\nu(H)$ takes into account the effect of variable depth. It has been numerically detected and experimentally confirmed that the soliton (KdV), upon transition to a smaller depth, decays into several solitons of different sizes in an oscillating trace.

He interpreted the breaking of waves over an inclined bottom as a consequence of the concentration of energy in a decreasing volume of water. In the process of wave transformation, the orbital velocities increase and the upper part of the crests become commensurate with the phase velocity of the wave. Due to the growth of the higher harmonics and the displacement of their phases relative to the main oscillation, the crest is sharpened and becomes asymmetric. The curvature of the orbital trajectories of particles increases, as a result of which the upward accelerations become comparable with the acceleration of gravity. According to the studies of D. Peregrine [Peregrine, 1976, 1993], in a thin wave front region, vertical accelerations can reach 5 g. As a result, the wave becomes unstable and begins to break down.

Theoretical and experimental studies of the interaction of gravity waves are carried out in the article by O. M. Phillips [Phillips, 1966]. The results of experimental studies on the detection of resonance interaction between surface waves are compared. These results give an idea of the main features of the interaction. The maximum response at resonance and the linear growth of the amplitude depend on the extent of the interaction. The decrease in the width of the group of interacting waves varies with the length of the interaction and because of the dispersion of the resonance points.

Especially it would be desirable to note research in the field of the theory of shallow water by E. N. Pelinovsky and representatives of his school. For quite some time he has been engaged in mathematical modeling of wave processes for predicting natural disasters, investigating tsunami phenomena, run-up of waves, and nonlinear deformation of the wave profile in shallow water. The work of L. A. Ostrovsky and E. N. Pelinovsky [Ostrovsky, Pelinovsky, 1970] is devoted to the transformation of waves on the surface of a fluid of variable depth. A generalized KdV equation is presented for the case when the depth $h = h(x)$ depends only on one coordinate x (perpendicular to the shore), and some of its solutions in application to the transformation of nonlinear waves on the shore slope.

In the work of A. G. Kulikovsky and V. A. Reutova [Kulikovsky, 1980] investigated the problem of nonlinear waves over semi-infinite underwater troughs and ridges. They developed an asymptotic wave model of the propagation of surface gravity waves.

Transformation of the spectrum of wind waves during refraction is presented in the work of V. P. Krasitsky [Krasitsky, 1974]. The effect of wind on the spectral components of the surface wave in shallow sea conditions is considered. With prolonged action of the wind, the waves grow, the maximum of the spectral characteristics moves toward smaller wave numbers.

A lot of works of I. T. Selezov have been devoted to the modeling of wave and diffraction processes in continuous media. In [Selezov, 1989] an equation is derived for weakly nonlinear and weakly dispersive waves on the surface of a fluid of variable depth. With the amplitude $\xi = 0$, of the bottom displacement, the resulting homogeneous equation reduces to an equation of the KdV type, and it has a soliton solution. The velocity of wave propagation depends not only on the depth and elevation of the free surface, but also on the magnitude of the displacement of the bottom

$$c = \sqrt{g(h_0 + \eta_0 - \xi)}\,. \qquad (3.1.5)$$

In the framework of the long-wave theory, he also considers the problem of propagation of an unsteady surface wave in the ocean from the focal region of an underwater earthquake, forms a free surface at various instants of time.

The work of I. T. Selezov also [Selezov, 1982] studies the transformation of waves in the coastal zone of the shelf. We consider a model with the restriction limited to the smallness of the nonlinearity parameter $\varepsilon = a/H$.

The approximation constructed is a generalization of the Boussinesq model or the shallow water model. Such a model describes the run-up of waves on an inclined beach at large values of ε more accurately than the models of Korteweg-de Vries, Peregrine and shallow water. The work [Selezov, 1985] presents a nonlinear refractive model of the propagation of surface waves in a liquid.

In the works of S. A. Gabov [Gabov, 1979, 1988] discusses the application of the Whitham equation to the description of the wave of sea waves on a gently sloping shore. The process of running waves from the high seas to the shore is divided into three main stages of evolution (Figure 3.1.2):

- smooth waves, at this stage the waves are far from the shore, the depth is great, they have a smooth, sloping nature;
- waves of limiting amplitude, approaching the shore at a comparatively shallow depth, these smooth waves increase their amplitude reaching the limiting amplitude, and sharpen at the vertex;
- wave breaking, at this stage the waves are overturned.

Having reached shallow water, the waves of extreme amplitude begin to collapse, toppling over the movement. To describe the process of approach, we use the version of the Whitham equation:

$$\frac{\partial \eta}{\partial t} + c_0 \frac{\partial \eta}{\partial x} + \frac{3}{2}\sqrt{\frac{g}{H}} \times \eta \frac{\partial \eta}{\partial x} + \int_{-\infty}^{+\infty} K_H(x-s)\eta_s(s,t)ds = 0, \qquad (3.1.6)$$

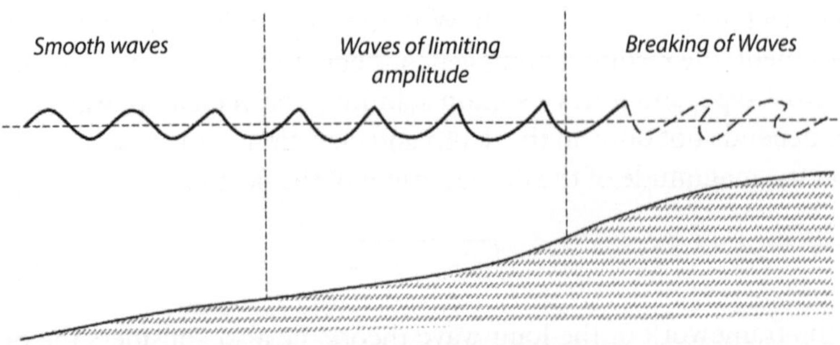

Figure 3.1.2 Stages of waves from the open seas to the shore.

the kernel of the integral term is defined by

$$K_H(x) = \frac{1}{2\pi}\int_{-\infty}^{+\infty} \sqrt{g\,\mathrm{th}(kH)/k} \times \exp(ikx)dk, \tag{3.1.7}$$

where $c(k) = \sqrt{g\,\mathrm{th}(kH)/k}$ – Phase velocity corresponding to the dispersion relation for waves on the surface of a heavy liquid with depth H.

It is assumed that the shore, on which the waves run, is sloping in the sense that at distances equal to several wavelengths, the depth varies insignificantly and can be considered *constant*. At distances d, characteristic of waves of a given stage of their evolution, the depth of the liquid varies little, i.e. $\Delta H/d << 1$. Therefore, the value H is assumed constant at each stage of wave evolution.

At the first stage, smooth waves are well described by the Whitham equation. The nonlinearity in the equation at this stage plays the role of only a factor restraining the dispersion effects, not allowing the waves to "*spread out*" due to different phase velocities of the harmonic components.

When moving to a shallower depth, the rear crests of the waves, being at a greater depth, have high propagation velocities (this follows from of the quantities $c_0 = \sqrt{gH}$ and $c(k) = \sqrt{g\,\mathrm{th}(kH)/k}$) and begin to catch up with the front crests. As a result, the wavelength decreases. However, the energy of waves concentrated at one wavelength remains unchanged due to the absence of dissipation of energy (or its insignificance), which leads to an increase in the amplitude of the waves.

Amplitudes of waves increase to a certain maximum level, and they turn into waves of limiting amplitude. The waves of the limiting amplitude, moving further into shallow water, begin to overturn and collapse. At small depths, the kernel of the integral term tends to zero almost everywhere. At the third stage of wave evolution, the action of the integral term is small (weak dispersion effect) and the solutions of the Whitham equation behave similarly to solutions of the simplest equations with nonlinearity, overturning and collapsing in a finite time. Questions on the overturning of waves on the basis of the Whitham equation are also described in the work of P. N. Naumkin, I. A. Shishmareva [Naumkin, 1986].

Some questions of the numerical modeling of the propagation of waves on the surface of a liquid in the framework of the nonlinear-dispersion Peregrine model are considered in the work of Yu. I. Shokin *et al.* [Shokin, 1989]. The paper also gives a classification of nonlinear dispersive models of shallow water.

Numerical simulation of nonstationary periodic surface waves is also carried out in the article by V. R. Kogan and V. V. Kuznetsov [Kogan, 1995]. The problem of the evolution of surface waves reduces to solving a system of ordinary differential equations. Dynamics of gravitational waves at various amplitudes and lengths of waves is numerically simulated. The profiles are of gravity waves at the initial moment of time and at the time of collapse.

Simulation of collapsing waves in a fluid of finite depth is described in the work of V. A. Kalmykov [Kalmykov, 1996], who uses a three-wave Fourier-wave-amplitude equation for this. Waves acquire a pronounced asymmetry in the direction of motion. The presence of asymmetry relatively the vertical axis was experimentally observed in the work of O. Phillips [Phillips, 1966], forming an asymmetric wave in the direction of travel we are given in M. Longuet-Higgins [Longuet-Higgins, 1993]. The problems of wave collapse are described in detail in a review by D. Peregrine [Peregrine, 1993].

The mathematical modeling of hydrodynamics of the Azov Sea is devoted to the work of L. A. Krukier, A. L. Chikin, G. V. Muratova and others. The work of L. A. Krukier [Krukier, 1991] is devoted to the assessment of the change in the hydrodynamics of the Sea of Azov in the case of the implementation of projects for the reconstruction of its ecosystem. For this, a mathematical model based on a numerical solution of the complete system of nonlinear shallow water equations is used. The papers [Chikin, 2001], [Muratova, 2004] describe the three-dimensional problem of determining the hydrodynamic flow parameters in the Azov Sea and the Taganrog Bay, and the flow patterns for various wind situations. The basis of these hydrodynamic models is the system of equations for the theory of shallow water with nonlinear terms, obtained in the work of N. E. Voltsinger and R. V. Piaskovsky [Voltsinger, 1977].

The construction of high-precision three-dimensional models of hydrodynamics of the Azov Sea is found in the work of A. I. Sukhinov. The works [Sukhinov, 1987, 2003, 2007] are devoted to the construction of high-precision 2D and 3D models of hydrodynamics, heat transport, salts and suspensions in shallow water. Physico-mathematical models of turbulent exchange in marine systems have been created, methods for solving grid equations of mathematical physics for mass parallel computing systems. The papers [Sukhinov, 2003], [Debol'skaya, Sukhinov, 2005] present the results of complex oceanological studies of the Sea of Azov, explain the mechanisms of the appearance of giant natural traps accumulating pollutants and suspended matter.

The article by A. A. Litvinenko and G. A. Khabakhpashev [Litvinenko, 1999] is devoted to numerical modeling of nonlinear long waves in basins with a shallow bottom. A nonlinear dispersion model of shallow water is considered taking into account topography and the viscosity of a liquid. A comparison is made of the calculations of the transformation of a plane disturbance of the free water surface with experimental data. The numerical problem is solved of the effect of a conic-cylindrical island, an underwater ridge and friction on an inclined bottom on the evolution of a flat solitary wave. The problems of modeling spatial nonlinear waves are considered in the work of D. G. Arkhipov and G. A. Khabakhpashev [Arkhipov, 2006]. The dynamics of finite amplitude waves running in different directions in the presence of dispersion.

A rather detailed review of nonlinear long waves on the surface of water and solitons was made in R. Kh. Zeytunyan [Zeytunyan, 1995]. In the review, a valid derivation of model equations using asymptotic methods is given. In some cases, these approximate equations are intuitive and heuristic. The historical moments of the investigation of waves on the surface of water are described, and the Boussinesq, Korteweg-de Vries and Schrödinger equations are considered. Various stages of the investigation of the phenomenon of solitary waves and solitons are presented.

Questions of the transformation of surface waves are considered in the works of Yu. Z. Aleshkov [Aleshkov, 1980, 1989, 2001]. He describes the problems of interaction of gravity waves with solid surfaces, dynamic processes in the coastal zone of the sea, the processes of transformation of surface waves in a liquid with a slowly varying depth, the nonlinear theory of transformation and refraction of surface waves.

The monograph of I. O. Leont'ev [Leont'ev, 2001] gives an idea of the main processes and mechanisms operating in the coastal zone of the seas, including the transformation of waves. The paper discusses a complex of phenomena caused by the collapse of wind waves, wave transport and horizontal circulation generated by the wave field. A numerical model of coastal circulation is constructed, which shows the features of storm currents for various types of coastal topography.

In the article by N. A. Kudryashov *et al.* [Kudryashov, 2003], mathematical modeling of gravity waves is carried out in the shallow-water approximation. The motion of a solitary wave along the water surface is studied, the wave exit to the shore, the passage of a solitary wave over an underwater rock. When modeling the motion of a soliton wave of soliton type, the approximation of shallow water begins to be violated at a ratio of the water depth to the wavelength equal to 0.3. The approximation of shallow water can be used to estimate the distance from the shore on which the wave is overturned.

In the works of S. Yu. Kuznetsov [Kuznetsov, 2002, 2005, 2009] investigations of surface waves in the coastal zone of the sea and transport of suspended sediments are conducted. The results of experiments on the measurement of waves are presented, the regularities of the transformation of irregular waves in the coastal zone. When the surface waves approach the shore, the wave heights are equalized because of the nonlinear rearrangement of the wave spectrum in the frequency range of the first harmonics. In this case, the ratio of the amplitudes of the first and second harmonics varies in time. The degree of asymmetry of the waves and the slope of the bottom lead to a frequency dependence of the dissipation rate of wave energy upon collapse. The collapse alters the shape of the wave spectrum, compensating for the effects of the nonlinear transformation process.

The work of M. A. Shermeneva and I. V. Shugan [Shermeneva, 2006] is devoted to mathematical modeling of the propagation of nonlinear waves on the water surface. It is a method for the analytical solution of the Boussinesq equations of high order. With the help of the solutions obtained, the run-up of surface waves on a sloping shore can be understood. Expressions describing the transformation of surface waves over an obstacle with elliptical level lines are derived. It is shown that the wave pattern on the surface can be used to judge the shape, location and orientation of the obstacle.

The works of E. N. Pelinovsky, I. I. Didenkoulova and A. A. Kurkina [Pelinovsky, 2006, 2008] investigate the processes of nonlinear deformation of the surface wave in shallow water and the problem of run-up of sea waves ashore within the framework of exact solutions of the nonlinear theory of shallow water. Depending on the shape of the suitable wave, different formulas are obtained for the height of the run-up of waves on shore. An analytical formula is proposed for the height of a single wave surge on shore in application to tsunami waves.

3.2 Nonlinear Models of Surface Gravity Waves in Shallow Water

Consider a fluid layer bounded by a horizontal bottom $z = 0$ and a free surface $z = H$ [Gabov, 1988]. Nonlinear waves on the surface of an ideal incompressible fluid in the field of gravity are described by a system of equations

$$\gamma \frac{\partial^2 \Phi}{\partial x^2} + \frac{\partial^2 \Phi}{\partial z^2} = 0 \text{ at } 0 < z < 1 + \varepsilon\zeta \,, \tag{3.2.1a}$$

$$\left. \frac{\partial \Phi}{\partial z} \right|_{z=0} = 0 \tag{3.2.1b}$$

$$\left.\begin{aligned}&\frac{\partial\zeta}{\partial t}+\varepsilon\frac{\partial\Phi}{\partial x}\frac{\partial\zeta}{\partial t}-\frac{1}{\gamma}\frac{\partial\Phi}{\partial z}=0\\&\frac{\partial\Phi}{\partial t}+\zeta+\frac{1}{2}\varepsilon\frac{\partial^2\Phi}{\partial x^2}+\frac{1}{2}\frac{\varepsilon}{\gamma}\frac{\partial^2\Phi}{\partial z^2}=0\end{aligned}\right\}\text{ at } z=1+\varepsilon\zeta\,, \tag{3.2.1c}$$

where $\gamma=(H/\lambda)^2$ – dispersion parameter, $\varepsilon=a/H$ – nonlinear parameter, a – amplitude of waves on a free surface, Φ – function of the potential of velocity, $\zeta(x,y,t)$ – function of a free surface.

Suppose that a function Φ, that is a solution of system (3.2.1) admits an expansion in a power series

$$\Phi=\sum_{n=0}^{\infty}\mathrm{f}_n(x,t)z^n\,. \tag{3.2.2}$$

Substituting this expansion in (3.2.1a) and the boundary condition (3.2.1b) and comparing the coefficients for various powers z, we obtain that (3.2.2) can be rewritten in the form

$$\Phi=\sum_{m=0}^{\infty}(-1)^m\gamma^m\frac{z^{2m}}{(2m)!}\frac{\partial^{2m}}{\partial x^{2m}}\mathrm{f}(x,t)\,, \tag{3.2.3}$$

where $\mathrm{f}(x,t)=\mathrm{f}_0(x,t)$.

Substituting (3.2.3) into the conditions (3.2.1c) on the free surface, we obtain

$$\frac{\partial\zeta}{\partial t}+\frac{\partial}{\partial x}\left[(1+\varepsilon\gamma)\mathrm{f}_x\right]-\left\{\frac{1}{6}(1+\varepsilon\gamma)^3\frac{\partial^4}{\partial x^4}\mathrm{f}+\frac{1}{2}\varepsilon(1+\varepsilon\gamma)^2\frac{\partial\gamma}{\partial x}\frac{\partial^3}{\partial x^3}\mathrm{f}\right\}\gamma$$
$$+O(\gamma^2)\zeta+\frac{\partial}{\partial t}\mathrm{f}+\frac{1}{2}\varepsilon\frac{\partial}{\partial x}\mathrm{f}^2-\frac{1}{2}(1+\varepsilon\gamma)^2$$
$$\left\{\frac{\partial^3\mathrm{f}}{\partial x^2\partial t}+\varepsilon\frac{\partial\mathrm{f}}{\partial x}\frac{\partial^3\mathrm{f}}{\partial x^3}-\varepsilon\frac{\partial^2\mathrm{f}^2}{\partial x^2}\right\}+O(\gamma^2). \tag{3.2.4}$$

Here the symbol $O(\gamma^2)$ denotes terms containing γ in the second degree and higher.

Suppose that, $\gamma=\left(H/\lambda\right)^2<<1$, that is the depth of the liquid is much smaller than the length of the waves propagating in the liquid. The depth of the liquid is much smaller than the length of the waves propagating in the liquid. Neglecting terms of the first and higher orders in (3.2.4) by γ, we obtain

$$\begin{aligned} &\frac{\partial \zeta}{\partial t}+\frac{\partial}{\partial x}\left[(1+\varepsilon\zeta)\frac{\partial \mathrm{f}}{\partial x}\right]=0 \\ &\zeta+\frac{\partial \mathrm{f}}{\partial t}+\frac{1}{2}\varepsilon\frac{\partial \mathrm{f}^2}{\partial x}=0 \end{aligned} \tag{3.2.5}$$

It follows from (3.2.3) that in the approximation under consideration $\Phi = \mathrm{f}(x,t)$ and thus

$$\frac{\partial \mathrm{f}}{\partial x}=\frac{\partial \Phi}{\partial x}=u,$$

where u – is the horizontal component of the particle velocity vector of the liquid. Differentiating the second equation in (3.2.5) with respect to x and returning to the dimensional variables, we obtain the following system of equations:

$$\begin{aligned} &\frac{\partial \zeta}{\partial t}+\frac{\partial}{\partial x}\left[(H+\zeta)u\right]=0 \\ &\frac{\partial u}{\partial t}+u\frac{\partial u}{\partial x}+g\frac{\partial \zeta}{\partial x}=0, \end{aligned} \tag{3.2.6}$$

which is called the *shallow water equations*.

The first equation in (3.2.6) in terms of its physical meaning is the equation of continuity or the equation of the law of conservation of the volume of a fluid. The second equation is dynamic and reflects the law of conservation of momentum.

The dispersion effects in these equations, in the approximation of shallow water (or longwave analysis) are absent. The solutions of these equations lead to waves with an increase in the level from time, and, consequently, they overturned.

The phenomenon of overturning is predicted by the theory of shallow water for all waves with an increase in the level. However, in reality, not all waves are overturned in shallow water. The inevitable overturning is the result of neglecting the dispersion effects. But it should be noted that simple theories that take into account dispersion effects in general lead to exclusion of the rollover process during propagation [Whitham, 1976].

To take into account dispersion effects, it is necessary to use the small parameter γ expansion for the equations of the theory of shallow water. In this approximation, the dispersion relation has the form

$$\omega^2 = c^2k^2 - \frac{1}{3}c^2H^2k^4 . \qquad (3.2.7)$$

Taking into account the relation (3.2.7), the system of shallow water equations (3.2.6) takes the form

$$\begin{gathered}\frac{\partial \zeta}{\partial t} + \frac{\partial}{\partial x}\left[(H+\zeta)u\right] = 0 \\ \frac{\partial u}{\partial t} + u\frac{\partial u}{\partial x} + g\frac{\partial \zeta}{\partial x} + \frac{1}{3}H\frac{\partial^3 \zeta}{\partial x \partial t^2} = 0.\end{gathered} \qquad (3.2.8)$$

These equations were first obtained by Boussinesq in 1871 and are named after him [Boussinesq, 1871]. This system describes nonlinear waves of small amplitude in shallow water.

Boussinesq's equations include waves moving both to the left and to the right because of the dispersion relation, when the frequency can be imaginary and real

$$\omega^2 = gk \, \mathrm{th}\, kH .$$

This relation was first obtained by Laplace; for further transformations we expand the dispersion relation for surface gravity waves (2.3.8) in a series

$$\omega = \sqrt{gk}\sqrt{\mathrm{th}\, kH} . \qquad (3.2.9)$$

Taking into account the hyperbolic tangent series expansion up to the second term [Dwight, 1983]

$$\mathrm{th}\, kH \approx kH - \frac{(kH)^3}{3}$$

from the expression (3.2.9) we obtain

$$\omega = \sqrt{gk}\sqrt{kH - \frac{(kH)^3}{3}} = \sqrt{gk}\sqrt{kH}\sqrt{1 - \frac{(kH)^2}{3}} . \qquad (3.2.10)$$

Further, we decompose the third root of the expression (3.2.10) into a power series in the parameter $kH << 1$ up to the second term

$$\sqrt{1-\frac{(kH)^2}{3}}=1-\frac{1}{2}\frac{(kH)^2}{3}. \tag{3.2.11}$$

We substitute the obtained expression (3.2.11) into the expression (3.2.10)

$$\begin{aligned}\omega &= \sqrt{gk}\sqrt{kH}\left(1-\frac{(kH)^2}{6}\right)=k\sqrt{gH}\left(1-\frac{(kH)^2}{6}\right)\\ &=\sqrt{gH}\cdot k-\sqrt{gH}\frac{H^2}{6}k^3=ck-c\frac{H^2}{6}k^3=ck-\beta k^3\end{aligned} \tag{3.2.12}$$

where

$$\beta=\frac{cH^2}{6},\ c=\sqrt{gH}$$

velocity of propagation of gravity waves for shallow water (expression (2.3.17)).

In the final form, the dispersion relation for waves moving to the right will take the form

$$\omega=ck-\beta k^3 \tag{3.2.13}$$

If we consider solutions of the Boussinesq equations describing waves propagating only to the right on the basis of the dispersion relation (3.2.13), then we obtain the *Korteweg-de Vries equation*

$$\frac{\partial\zeta}{\partial t}+\frac{\partial\zeta}{\partial x}+\varepsilon\zeta\frac{\partial\zeta}{\partial x}+\frac{1}{6}\gamma\frac{\partial^3\zeta}{\partial x^3}=0. \tag{3.2.14}$$

It was derived in 1885 to describe the propagation of waves in channels [Gabov, 1988]. Taking into account the values of dimensionless parameters, equation (3.2.14) takes the form

$$\frac{\partial\zeta}{\partial t}+c\left(1+\frac{3}{2}\frac{\zeta}{H}\right)\frac{\partial\zeta}{\partial x}+\frac{1}{6}cH^2\frac{\partial^3\zeta}{\partial x^3}=0. \tag{3.2.15}$$

The solutions of the Boussinesq equation system (3.2.8) satisfy the relation between the horizontal component of the particle velocity u and the function of the perturbed surface ζ (the Riemann invariant [Whitham, 1976])

$$u = \zeta - \frac{1}{4}\varepsilon\zeta^2 + \frac{1}{3}\gamma\frac{\partial^2\zeta}{\partial x^2}$$

Then the Korteweg-de Vries equation for the horizontal component of the particle velocity u takes the following form

$$\frac{\partial u}{\partial t} + c\frac{\partial u}{\partial x} + \beta\frac{\partial^3\zeta}{\partial x^3} = -\varepsilon u\frac{\partial u}{\partial x}. \qquad (3.2.16)$$

The Korteweg-de Vries equation is applied to many problems on dispersive waves that are not associated with waves on water: ion-sound waves in a cold plasma, longitudinal waves in elastic rods [Nonlinear Wave, 1977].

Also often used is the canonical form of the Korteweg-de Vries equation, which is distinguished by the simplicity of the notation

$$u_t - 6uu_x + u_{xxx} = 0. \qquad (3.2.17)$$

The term on the right-hand side of (3.2.16) describes nonlinear processes, and the third term on the left-hand side is the dispersion of the waves. The second and third terms on the left-hand side are derived from the dispersion relation (2.3.8). In this case, a weak nonlinearity $\varepsilon << 1$, leading to wave steepening and a weak dispersion $\gamma << 1$, leading to its blurring can compensate each other. In this case, a stationary nonlinear wave propagating without changing the shape with a constant velocity will arise.

The periodic solution of the Korteweg-de Vries equation with the help of Jacobi's elliptic function is [Whitham, 1976]

$$u(x,t) = \varepsilon \times cn^2\left[\sqrt{\frac{3\gamma}{4H^2}} \times (x - Ut)\right], \qquad (3.2.18)$$

where $U = c(1-\gamma/2)$ - wave propagation velocity, $cn(z) = \cos(\phi) = \sqrt{1-x^2}$ – Jacobi elliptic function [Dwight, 1983], [Abramovitz, 1979].

The modulus m of the elliptic function is equal to $m=\sqrt{\varepsilon/\gamma}$, and the wavelength,

$$\lambda = \frac{4H}{\sqrt{3\gamma}} \times K(m),$$

where $K(m)$ - complete Legendre elliptic integral of the first kind

$$K(m) = \int_0^{\pi/2} \frac{d\phi}{\sqrt{1-m^2 \sin^2 \phi}}.$$

Because of Jacobi's elliptic function $cn(z)$, Korteweg and de Vries called these solutions *cnoidal waves.* The module m gives a comparative estimate of the effects of nonlinearity and dispersion. In the linear limit $m \to 0$ in expression (3.2.18), $cn(x) \to \cos x$, at $m \to 1$, $cn(x) \to \mathrm{sech}(x)$ - corresponds to a solitary wave (soliton).

The solutions of the Korteweg-de Vries equation are correct for values $0 \le \varepsilon \le \gamma$, when ε and γ are small. The cnoidal waves are limited in height and are in fact a periodic sequence of solitons in shape (Figure 3.2.1).

The balance between non-linearity and variance is often specified using the Ursell criterion (number):

$$Ur = \frac{3\varepsilon}{4\gamma}. \tag{3.2.19}$$

A dimensionless parameter *Ur* first appeared in the work of J. Stokes (Stokes, 1849). However, in more detail, as a measure of the ratio of non-linearity to dispersion, F. Ursell [Ursell, 1966].

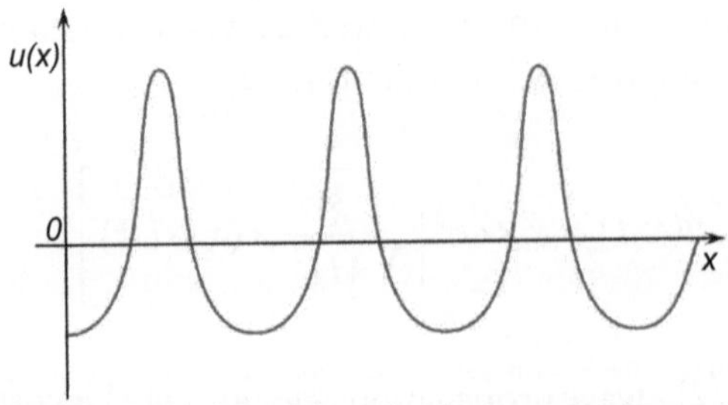

Figure 3.2.1 Cnoidal waves.

3.3 Solution of the Nonlinear Shallow Water Equation by the Method of Successive Approximations

The invariance of the forms of cnoidal waves does not allow us to trace the dynamics of nonlinear surface gravity waves in the course of propagation in shallow water.

To follow the dynamics of the process of propagation of surface gravity waves, we consider the shallow water equation (from the expression (3.2.6)) with allowance for the quadratic nonlinearity:

$$\frac{\partial u}{\partial t}+c\frac{\partial u}{\partial x}=-\varepsilon u\frac{\partial u}{\partial x} \tag{3.3.1}$$

Equation (3.3.1) is a part of the Korteweg-de Vries equation (3.2.16), disregarding wave dispersion. As long as the dispersion is neglected, in the subsequent considerations it will be taken into account in its original form in terms of the known dispersion law in Eq. (3.2.13). It can be noted that the effect of dispersion becomes more pronounced for the higher harmonics of the fundamental wave, which lead to a steepening of the wave profile.

As is known, the classical and the most direct way of solving nonlinear equations is the solution by the method of successive approximations [Lamb, 1947], [Vinogradova, Rudenko, 1979], [Abbasov, Monograph, 2007]. Both the shallow water equation and the method of successive approximations stand at the origins of the creation of nonlinear wave phenomena. This equation and method in the middle of the nineteenth century described the tidal waves of G. Airy and waves of finite amplitude of J. Stokes. However, the process of generating higher harmonics, leading to a sharpening of the wave profile, does not always satisfy the law of conservation of energy. The increase in the higher harmonics does not take into account the decrease in the energy of the main wave.

When the shallow water conditions are satisfied $H/\lambda << 1$, when the horizontal particle velocity ϑ_x is $1/kH$ times larger than the vertical one ϑ_z the nonlinearity parameter ε is determined by the expression $\varepsilon = \vartheta_x/c = a/H$ [Brekhovskikh, 1982]. Let, at $t = 0$ be a perturbation

$$u(x,0) = \varepsilon U_0 \exp(ikx) + c.c.,$$

where $(c.c.)$ – denotes the term complex conjugate to the first one, which is introduced into nonlinear problems to take into account that u is real-valued.

Let us seek a solution of equation (3.3.1) with the method of successive approximations in the form of the expansion in terms of the small parameter ε (ε<<1), restrict ourselves to the first two terms

$$u = \varepsilon u^{(1)} + \varepsilon^2 u^{(2)}, \tag{3.3.2}$$

and we believe that $u^{(n+1)} << u^{(n)}$ [Vinogradova, Rudenko, 1979].

We substitute expression (3.3.2) into equation (3.3.1)

$$\frac{\partial(\varepsilon u^{(1)} + \varepsilon u^{(2)})}{\partial t} + c\frac{\partial(\varepsilon u^{(1)} + \varepsilon u^{(2)})}{\partial x} = -\varepsilon(\varepsilon u^{(1)} + \varepsilon u^{(2)})\frac{\partial(\varepsilon u^{(1)} + \varepsilon u^{(2)})}{\partial x}$$

For terms containing ε in the first approximation, we obtain the linear equation

$$\frac{\partial u^{(1)}}{\partial t} + c\frac{\partial u^{(1)}}{\partial x} = 0. \tag{3.3.3}$$

Solution of (3.3.3) with initial conditions $t = 0$,

$u^{(1)}(x,0) = U_0 \exp(ikx) + c.c.$, will be a function

$$u^{(1)}(x,t) = U_0 \exp\left[i(kx - \omega t)\right] + c.c. \tag{3.3.4}$$

Equating now the terms with ε^2 in the second approximation, for $u^{(2)}$ we obtain an inhomogeneous linear equation

$$\frac{\partial u^{(2)}}{\partial t} + c\frac{\partial u^{(2)}}{\partial x} = -\frac{\partial u^{(1)^2}}{2\partial x}. \tag{3.3.5}$$

The right-hand side of this equation is a solution of the linear equation (3.3.3) for $u^{(1)}$. To find the solution $u^{(2)}(x,t)$ we squared the expression (3.3.4) for $u^{(1)}(x,t)$

$$\begin{aligned} u^{(1)2}(x,t) &= \left(U_0 \exp\left[i(kx - \omega t)\right] + U_0 \exp\left[-i(kx - \omega t)\right]\right)^2 \\ &= U_0^2 \exp\left[2i(kx - \omega t)\right] + 2U_0^2 \exp\left[i(kx - \omega t) - i(kx - \omega t)\right] \\ &\quad + U_0^2 \exp\left[-2i(kx - \omega t)\right] \\ &= 2U_0^2 + U_0^2 \exp\left[2i(kx - \omega t)\right] + (c.c.) \end{aligned}$$

We differentiate the resulting expression by x

$$\frac{\partial u^{(1)2}(x,t)}{\partial t}$$
$$= \frac{\partial(2U_0^2)}{\partial x} + \frac{\partial\left[U_0^2 \exp\left[2i(kx-\omega t)\right]\right]}{\partial x} + \frac{\partial(U_0^2 \exp\left[-2i(kx-\omega t)\right])}{\partial x}$$
$$= 2ikU_0^2 \exp\left[2i(kx-\omega t)\right] + (c.c.)$$

Equation (3.3.5), taking into account the transformations takes the form

$$\frac{\partial u^{(2)}}{\partial t} + c\frac{\partial u^{(2)}}{\partial x} = -ikU_0^2 \exp\left[i(2kx-2\omega t)\right] + (c.c.)$$

The solution for $u^{(2)}(x,t)$ when no dispersion is present (with initial condition $u^{(2)}(x,0)=0$) has the following form

$$u^{(2)}(x,t) = -iktU_0^2 \exp\left[i(2kx-2\omega t)\right] + c.c. \tag{3.3.6}$$

Then the general solution for the horizontal particle velocity taking into account the first approximation (3.3.4) and the second (3.3.5) approximation takes the form

$$u(x,t) = \varepsilon u^{(1)}(x,t) + \varepsilon^2 u^{(2)}(x,t)$$
$$= \varepsilon U_0 \exp\left[i(kx-\omega t)\right] - ikt\varepsilon^2 U_0^2 \exp\left[i(2kx-2\omega t)\right] + c.c. \tag{3.3.7}$$

Analyzing expression (3.3.7), it can be noted that the amplitude of the fundamental harmonic remains constant, and the amplitude of the second harmonic increases linearly with time (because of the secular term). Therefore, in the absence of dispersion, for any arbitrarily small value of ε the instability (breaking) of the wave will occur, which is determined by equating the amplitudes of the terms $\varepsilon\, u^{(1)}$ and $\varepsilon^2 u^{(2)}$:

$$\varepsilon U_0 = \varepsilon^2 k t_{char.}^{(1)} U_0^2$$

hence, the time period after which instability occurs is

$$t_{char.}^{(1)} = 1/\varepsilon k U_0 \tag{3.3.8}$$

It should be noted that by this time the method of successive approximations will be unacceptable, since the primary condition is violated $u^{(n+1)} << u^{(n)}$.

However, up to this point, with the growth of the second harmonic, its interaction with the fundamental harmonic will begin to increase. This interaction will lead to the excitation of the third, fourth, etc. harmonics. Consequently, with time, higher frequencies will appear in the wave spectrum, which will lead to the transformation of the wave profile and to the steepening of its front.

In order to identify these changes, let us look for a solution of equation (3.3.1) by the method of successive approximations with new initial conditions

$$u_{sec.}(x,t) = \varepsilon u^{(1)}_{prim.}(x,t) + \varepsilon^2 u^{(2)}_{sec.}(x,t) \quad (3.3.9)$$

Then the expression for the horizontal velocity of the particles of the medium in the first approximation will have the form

$$u^{(1)}_{prim.}(x,t) = u(x,t) = \varepsilon U_0 \exp\left[ikx\right] - ikt\varepsilon^2 U_0^2 \exp\left[i2kx\right] + c.c., \quad (3.3.10)$$

(the time factors $\exp\left[i\omega t)\right]$ are omitted here and below).

The inhomogeneous linear equation for the new initial conditions will have the form

$$\frac{\partial u^{(2)}_{sec.}}{\partial t} + c\frac{\partial u^{(2)}_{sec.}}{\partial x} = -\frac{\partial u^{(1)^2}_{prim.}}{2\partial x}. \quad (3.3.11)$$

Let us square the solution for the first approximation (3.3.10), for convenience we denote $\varepsilon U_0 = A_1$, $-ikt\varepsilon^2 U_0^2 = A_2$

$$\begin{aligned}
&u^{(1)^2}_{prim.}(x,t)\\
&= \left[A_1 \exp\left[ikx\right] + A_2 \exp\left[i2kx\right] + A_1 \exp\left[-ikx\right] + A_2 \exp\left[-i2kx\right]\right]^2\\
&= A_1^2 \exp\left[i2kx\right] + A_1 A_2 \exp\left[ikx + i2kx\right] + A_1^2 \exp\left[0\right]\\
&\quad + A_1 A_2 \exp\left[ikx - i2kx\right] + A_2 A_1 \exp\left[i2kx + ikx\right] + A_2^2 \exp\left[i4kx\right]\\
&\quad + A_2 A_1 \exp\left[i2kx - ikx\right] + A_2^2 \exp\left[0\right] + A_1^2 \exp\left[0\right]\\
&\quad + A_1 A_2 \exp\left[-ikx + i2kx\right] + A_1^2 \exp\left[-i2kx\right]\\
&\quad + A_1 A_2 \exp\left[-ikx - i2kx\right] + A_2 A_1 \exp\left[-i2kx + ikx\right]
\end{aligned}$$

$$+A_2^2\exp[0]+A_2A_1\exp[-i2kx-ikx]+A_2^2\exp[-i4kx]$$
$$=2A_1^2+2A_2^2+\left(A_1^2\exp[i2kx]+(c.c.)\right)+\left(A_2^2\exp[i4kx]+(c.c.)\right)+$$
$$+\left(2A_1A_2\exp[i3kx]+(c.c.)\right)+\left(2A_1A_2\exp[ikx]+(c.c.)\right).$$

(3.3.12)

We differentiate the expression (3.3.12) along the coordinate x

$$\frac{\partial u_{prim.}^{(1)^2}}{\partial x}=\left[i2kA_1^2\exp[i2kx]+i4kA_2^2\exp[i4kx]+\right.$$
$$\left.+i3k\cdot 2A_1A_2\exp[i3kx]+ik\cdot 2A_1A_2\exp[ikx]\right]+(c.c.) \quad (3.3.13)$$

We substitute values of parameters A_1, A_2

$$\frac{\partial u_{prim.}^{(1)^2}}{\partial x}=\left[i2k\cdot\varepsilon^2U_0^2\exp[i2kx]+i4k\cdot(-k^2t^2\varepsilon^4U_0^4)\exp[i4kx]+\right.$$
$$+i3k\cdot 2\cdot\varepsilon U_0\cdot(-ikt\varepsilon^2U_0^2)\exp[i3kx]$$
$$\left.+ik\cdot 2\cdot\varepsilon U_0\cdot(-ikt\varepsilon^2U_0^2)\exp[ikx]\right]+(c.c.)$$
$$=\left[2ik\varepsilon^2U_0^2\exp[i2kx]-4ik^3t^2\varepsilon^4U_0^4\exp[i4kx]+\right.$$
$$\left.+6k^2t\varepsilon^3U_0^3\exp[i3kx]+2k^2t\varepsilon^3U_0^3)\exp[ikx]\right]+(c.c.)$$

(3.3.14)

The solution for the second approximation takes the form

$$u_{sec.}^{(2)}(x,t)=\left[ikt\varepsilon^3U_0^2\exp(2ikx)-2ik^3t^3\varepsilon^4U_0^4\exp(4ikx)+\right.$$
$$\left.+3k^2t^2\varepsilon^3U_0^3\exp(3ikx)+k^2t^2\varepsilon^3U_0^3\exp(ikx)\right]+(c.c.) \quad (3.3.15)$$

Analyzing expression (3.3.15), it can be noted that as a result of the interaction of the main wave k with the second $2k$, secondary waves arise not only with doubled values of $2k$ and $4k$, but also the combination waves k and $3k$. Accordingly, in the spectrum of the initial wave, the second, third and fourth harmonics appear with time, which will lead to distortion of the wave profile. Here, it should be emphasized that a wave with a frequency of the fundamental wave appears in a set of secondary waves, and this wave has a quadratic dependence on time.

The final solution in two approximations for the horizontal velocity of the particles of the medium after substituting the expressions (3.3.10) and (3.3.15) into expression (3.3.9) we obtain

$$\begin{aligned} u_{sec.}(x,t) = & \left[\varepsilon^2 U_0 \exp(ikx) - ikt\varepsilon^3 U_0^2 \exp(2ikx)\right] + \\ & + \left[k^2 t^2 \varepsilon^5 U_0^3 \exp(ikx) + ikt\varepsilon^4 U_0^2 \exp(2ikx) + \right. \\ & \left. + 3k^2 t^2 \varepsilon^5 U_0^3 \exp(3ikx) - 2ik^3 t^3 \varepsilon^6 U_0^4 \exp(4ikx)\right] \\ & + (c.c.) \end{aligned} \tag{3.3.16}$$

According to the expression (3.3.16) obtained after a certain time interval, as a result of the rapid growth of the amplitudes of the second term of expression (3.3.9), the wave instability will occur, since the second term contains higher degrees of time dependence. This characteristic time can also be determined by equating the terms

$$\varepsilon u_{prim.}^{(1)}(x, t_{char.}^{(2)}) = \varepsilon^2 u_{sec.}^{(2)}(x, t_{char.}^{(2)}), \tag{3.3.17}$$

i.e., when the sum of the amplitudes of the secondary wave becomes equal to the sum of the amplitudes of the primary. But, unlike expression (3.3.8), the dependence here is more complicated, so an analytical definition of the expression for the characteristic time $t_{char.}^{(2)}$ for a given interaction will be quite laborious, so we will use the graphical method in further studies.

3.4 Modeling the Propagation of Nonlinear Surface Gravity Waves in Shallow Water

To study nonlinear shallow-water wave effects, we will use the hydrologic conditions of the Taganrog Bay of the Azov Sea as a model. The mean depth of the bay does not exceed 5 m [Mamykina, 1980], therefore, the shallow water conditions will be satisfied for gravity waves with their lengths longer than 30 m. The bay floor is flat, and surface tension is absent. Swell waves are regarded as the initial surface waves; i.e., the effect of wind is not taken into account.

As is known, eddy-free surface gravity waves overturn at a height to length ratio of $2a/\lambda = 0{,}141$ [Lamb, 1947], [Sretensky, 1977], although the real waves are also upset with less steepness [Benjamin, 1967]. Therefore, we use the initial values of the steepness in the range from 0.01 to 0.05.

To follow the energy parameters of gravity waves, we use expression (2.3.23):

$$E_n = \frac{\rho_0 g}{4}\lambda_n a_n^2 = \frac{\rho_0 g}{4}\lambda_n\left(\frac{U_{n0}k_n H}{\omega_n}\right)^2, \tag{3.4.1}$$

(taking into account relation $a_n = (U_{n0}k_n H/\omega_n)$), where ρ_0 – fluid density, n – number of the corresponding harmonic n = 1,2,3,4.

Let us observe variations in the profile of gravity wave after it enters the bay; its initial parameters are the following: frequency f = 0.2 Hz; length λ = 35 m; propagation velocity c = 7 m/s; initial steepness $-2a/\lambda$ = 0,024; a = 0,413 m; kH = 0,89; ε = 0,083.

It is necessary to determine the spatial (time – $t^{(2)}_{char.}$) interval $x^{(2)}_{char.}$. This distance is found graphically based on the expression (3.3.17), in our case for a gravity wave with the given initial conditions it is $x^{(2)}_{char.} \approx 10$ km.

But before proceeding to the calculations, we should note some physical features of the wave processes that occur. In expression (3.3.16), the amplitudes of all harmonics (except the fundamental) increase with time. However, the amplitude of the fundamental harmonic remains constant in this case. Although higher harmonics are energetically provided from the fundamental harmonic.

Consequently, the amplitude of the primary waves in expression (3.3.16) should decrease with time. Therefore, to study the distortions of the gravity wave profile, it is necessary to take into account these physical features of the wave processes (otherwise, this may result in the violation of the law of energy).

As is known, in the absence of dissipation, the energies of the primary and secondary waves must correspond to the energy conservation law

$$E_1^{(1)}(x) + E_2^{(1)}(x) = E_1^{(2)}(x) + E_2^{(2)}(x) + E_3^{(2)}(x) + E_4^{(2)}(x), \tag{3.4.2}$$

where $E_n^{(1)}(x)$, $E_n^{(2)}(x)$ – energy respectively primary and secondary waves.

The law of decreasing amplitude of primary waves must satisfy the following condition

$$E^{(1)}_{prim.}(x) = E - E^{(2)}_{sec.}(x), \tag{3.4.3}$$

where $E^{(1)}_{prim.}(x)$, $E^{(2)}_{sec.}(x)$ – total energy of primary and secondary waves, E – The fundamental energy from the expression (3.4.1):

$$E = \frac{\rho_0 g}{4}\frac{2\pi}{k}\left(\frac{U_0 kH}{\omega}\right)^2 = \frac{\pi\rho_0 g}{2\omega}\left(\frac{k}{\omega}\right)\left(U_0 H\right)^2 = \left(\frac{\pi\rho_0 g}{2\omega c}\right)(U_0 H)^2, \quad (3.4.4)$$

also for the amplitude of the velocity $U_0 = \left(\omega a / k H\right)$, he amplitude of the displacements of the free surface a is determined from the expression for the steepness – $2a/\lambda$.

To find the amplitude of the horizontal velocity for a known energy, we use expression (3.4.4) (by inverse dependence):

$$U^{(1)}_{prim.} = \frac{1}{H}\sqrt{\frac{2\omega c}{\rho_0 g \pi}\cdot E^{(1)}_{prim.}}. \quad (3.4.5)$$

However, this value of the amplitude of the horizontal velocity is the sum of the amplitudes of the fundamental and second harmonics (expression (3.3.10)). In this case, the law of decreasing amplitude of the fundamental harmonic is found on the basis of expression (3.3.16), which contains the time dependence to the third power. Then we find the amplitude of the second harmonic in terms of the difference between the total amplitude of the primary waves and the amplitude of the fundamental harmonic:

$$U^{(1)}_2 = U^{(1)}_{prim.} - U^{(1)}_1. \quad (3.4.6)$$

Figure 3.4.1a shows the graphs of the velocities of the primary $u^{(1)}_{prim.}(x,t)$ and secondary waves $u^{(2)}_{sec.}(x,t)$, as well as the fundamental $u^{(1)}_1(x,t)$ and second harmonics $u^{(1)}_2(x,t)$. For convenience, they are presented depending on the distance x.

As already noted, instability occurs at a distance of $x^{(2)}_{char.} \approx 10$ km (i.e., in $t^{(2)}_{char.} \approx 24$ min.). However, our model is correct up to a distance of $x \approx 6$ km, i.e. as long as the primary condition is fulfilled $u^{(n+1)} << u^{(n)}$, when the value for the second approximation is an order of magnitude smaller than the value for the first approximation.

On the basis of the analysis in Figure 3.4.1a, it can be noted that with an increase in the total amplitude of the secondary waves, the amplitude of the primary waves and the fundamental harmonic fall. And the amplitude of the second harmonic slowly increases, and at the instant of instability falls

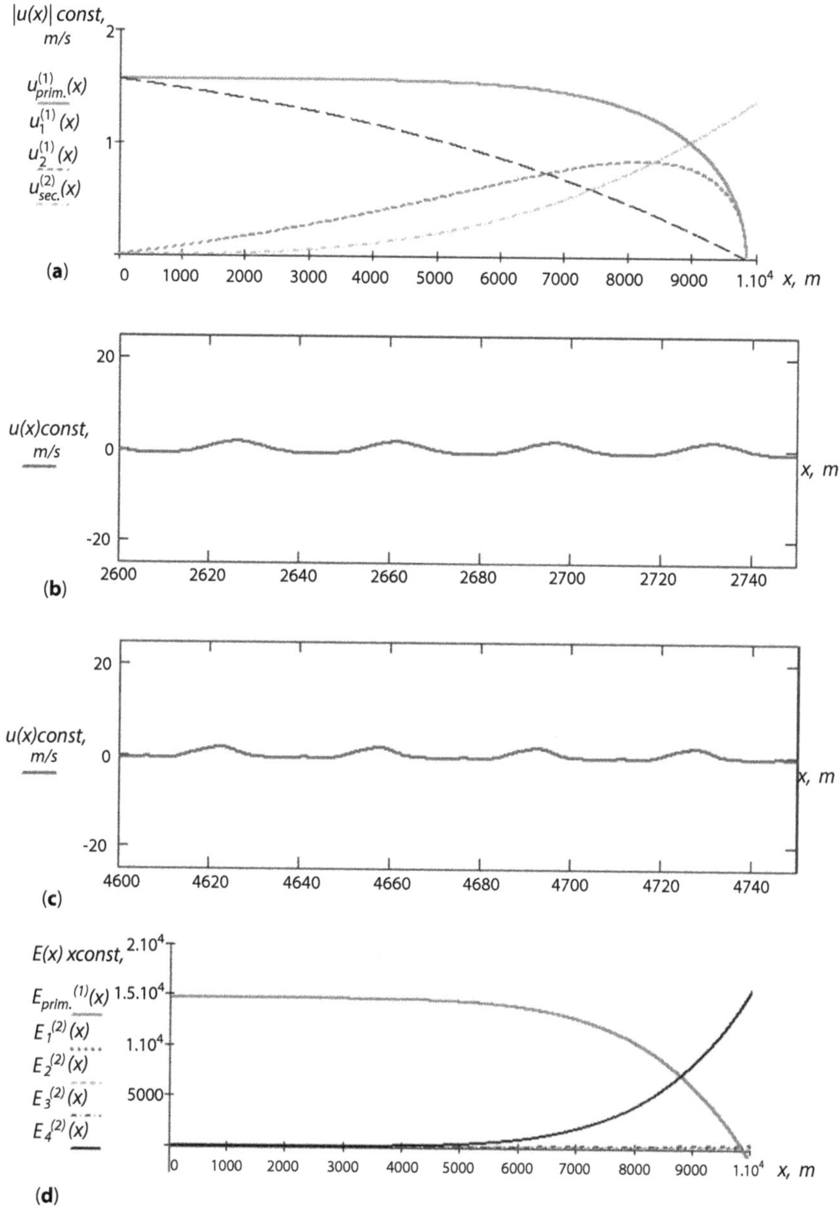

Figure 3.4.1 The main characteristics of a surface gravity wave with parameters: $f = 0.2$ Hz; $\lambda = 35$ m; $c = 7$ m/s; H = 5 m; $2a/\lambda = 0.024$; $kH = 0.89$; $\varepsilon = 0.083$; $\gamma = 0.02$.

sharply. Thus, the law of conservation of energy for a given wave process is fully observed.

Consider the change in the profile of the initial gravity wave to a distance of $x \approx 6$ km. Figures 3.4.1b and 3.4.1b show the dependence of the variation of the horizontal velocity $u(x,t)$ on the mean free path with the expression (3.3.16) taken into account, the vertical scale is increased.

The wave with initially cosine profile during the run is gradually distorted, the crests are sharpened, and the troughs become more flat. This is the result of the increasing influence of high-frequency harmonics. The collapse of such a wave can occur in the form of a breaker, slipping off the crest [Monin, 1985]. The reason for the breaking in our case is naturally the depth of the fluid.

The energy characteristics of the wave process are shown in Figure 3.4.1e, where the energy of the primary and secondary waves is plotted as a function of the distance x. The energy of the primary waves $E^{(1)}_{prim.}(x)$ decreases by the instability moment, respectively, the energy of the secondary waves $E^{(2)}_1(x)$, $E^{(2)}_2(x)$, $E^{(2)}_3(x)$ and $E^{(2)}_4(x)$ grows. In this case, a complete transfer of energy from primary waves to secondary waves occurs. According to the graph, the energy of the fourth harmonic grows faster than all – $E^{(2)}_4(x)$, which has the greatest degree of dependence on time.

Figure 3.4.2 shows the growth curves of the amplitudes of secondary wave velocities from distance. Here it is quite clearly visible the relation between harmonics.

Among the secondary waves, the amplitude of the velocity of the fourth harmonic, the third, the first and the second, grow fastest of all. This regularity is also observed for gravity waves with other initial conditions.

Figure 3.4.3a – 3.4.3e presents the main characteristics of another surface wave with parameters: frequency $f = 0.09$ Hz; wavelength $\lambda = 77.8$ m;

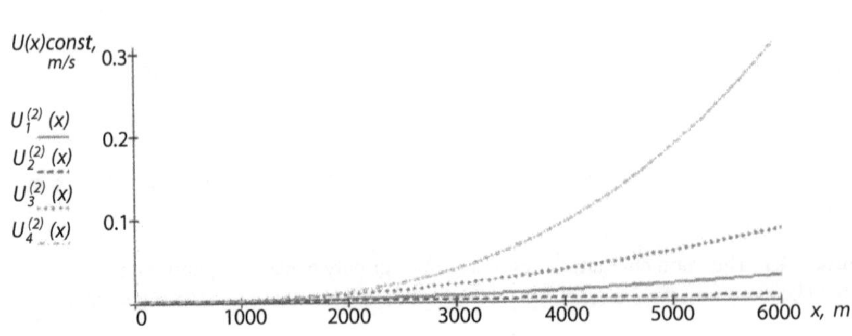

Figure 3.4.2 Accumulation of amplitudes of secondary waves of a surface gravity wave with parameters: $f = 0.2$ Hz; $\lambda = 35$ m; H = 5 m; $2a/\lambda = 0.024$; $kH = 0.89$; $\varepsilon = 0.083$; $\gamma = 0.02$.

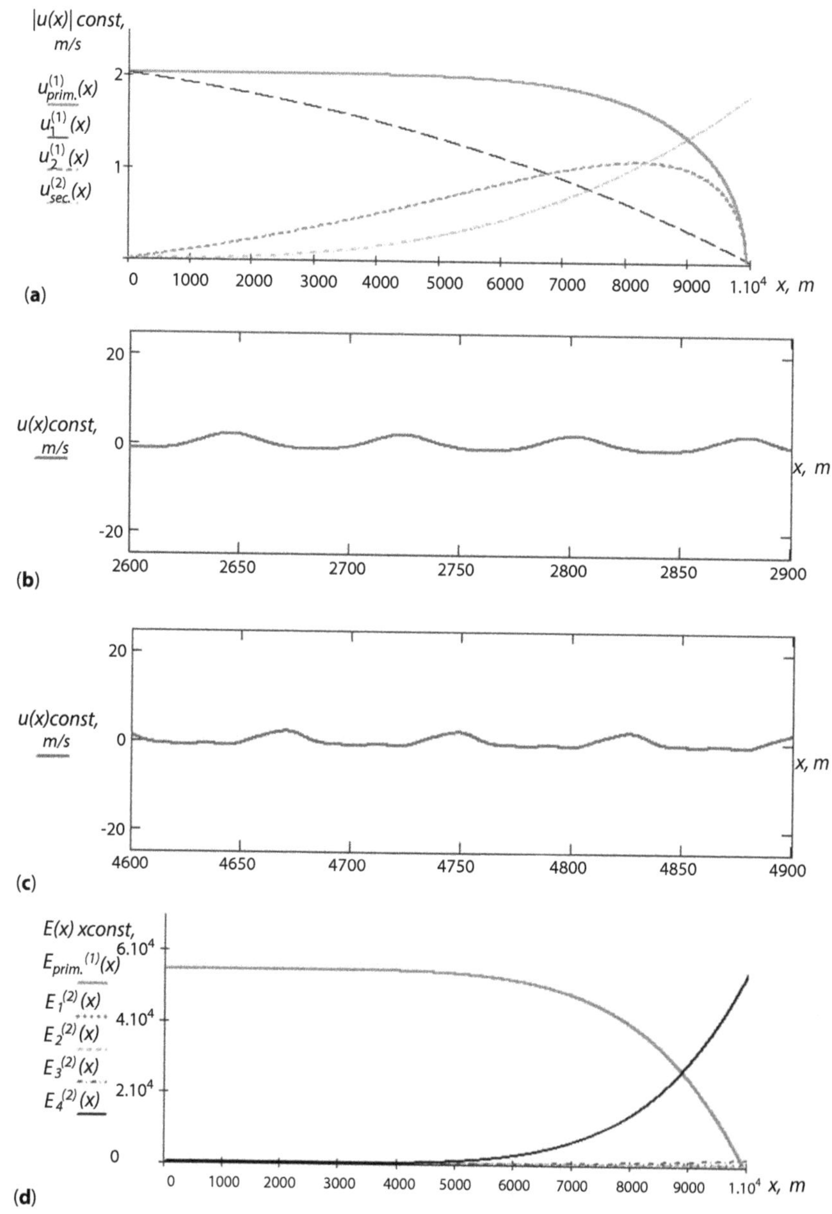

Figure 3.4.3 Main characteristics of the surface gravity waves with parameters: $f = 0.09$ Hz; $\lambda = 77.8$ m; $c = 7$ m/s; H = 5 m; $2a/\lambda = 0.014$; $kH = 0.4$; $\varepsilon = 0.107$; $\gamma = 0.004$.

the initial slope is – $2a/\lambda = 0.014$; $a = 0.537$ m; $kH = 0.4$. According to Figure 3.4.3a, for this wave the instability also occurs at a distance of – $x^{(2)}_{char.} \approx 10$ km. Despite the fact that this wave has a lower steepness than the previous one, it, due to the longer length, also comes to instability at the same distance.

Figure 3.4.4a – 3.4.4e shows the main characteristics of the same wave with a large initial steepness – $2a/\lambda = 0.015$, which increases the nonlinear parameter – $\varepsilon = 0.118$. This leads to a decrease in the instability distance, which already comes at a distance of – $x^{(2)}_{char.} \approx 7.5$ km, and our model is correct to a distance of $x \approx 4$ km.

With an increase in the initial steepness, the instability distance decreases, with increasing wave frequency (at constant steepness) this distance increases, as the value of the nonlinear parameter decreases.

Figure 3.4.5 shows three-dimensional spatial models of the gravity wave propagation along the bay, the main characteristics of which were shown in Figure 3.4.3: $\lambda = 77.8$ m, $2a/\lambda = 0{,}014$. For the calculations we used a section of the bay with the following parameters: depth $H = 5$ m; the length along the axis is $x = 10$ km; on the axis $y = 100$ m, grid spacing is 3.5 m; scales along the y and z axes are increased.

At three propagation sites, the gradual distortion of the profile of the gravity wave, whose crests are becoming steeper as it spreads, is clearly demonstrated.

Figure 3.4.6 shows the three-dimensional spatial models of the gravity wave propagation along the bay with the following parameters: $\lambda =$ 155.6 m; $2a/\lambda = 0.0088$. This wave has twice the length (relative to the previous wave), the instability distance is also $x^{(2)}_{char.} \approx 10$ km.

It should be emphasized that for the emerging higher harmonics the shallow water condition may not be observed. And this can lead to a dispersion of the propagation velocity, which can cause inhibition of nonlinear processes [Whitham, 1976]. But when approaching the shore due to the decrease in depth, the shallow water condition will be satisfied better and, consequently, the wave breaking.

Steepening of the Leading Front of the Wave

Analyzing the constructed profiles of surface gravitational waves, it is necessary to note the following features. From full-scale observations it is known that the profile of real surface waves is asymmetric: the crests are relatively steep and short, pointed, and the troughs are flattened and broad [Lamb, 1947]. Orbital movements of particles are also asymmetric: their

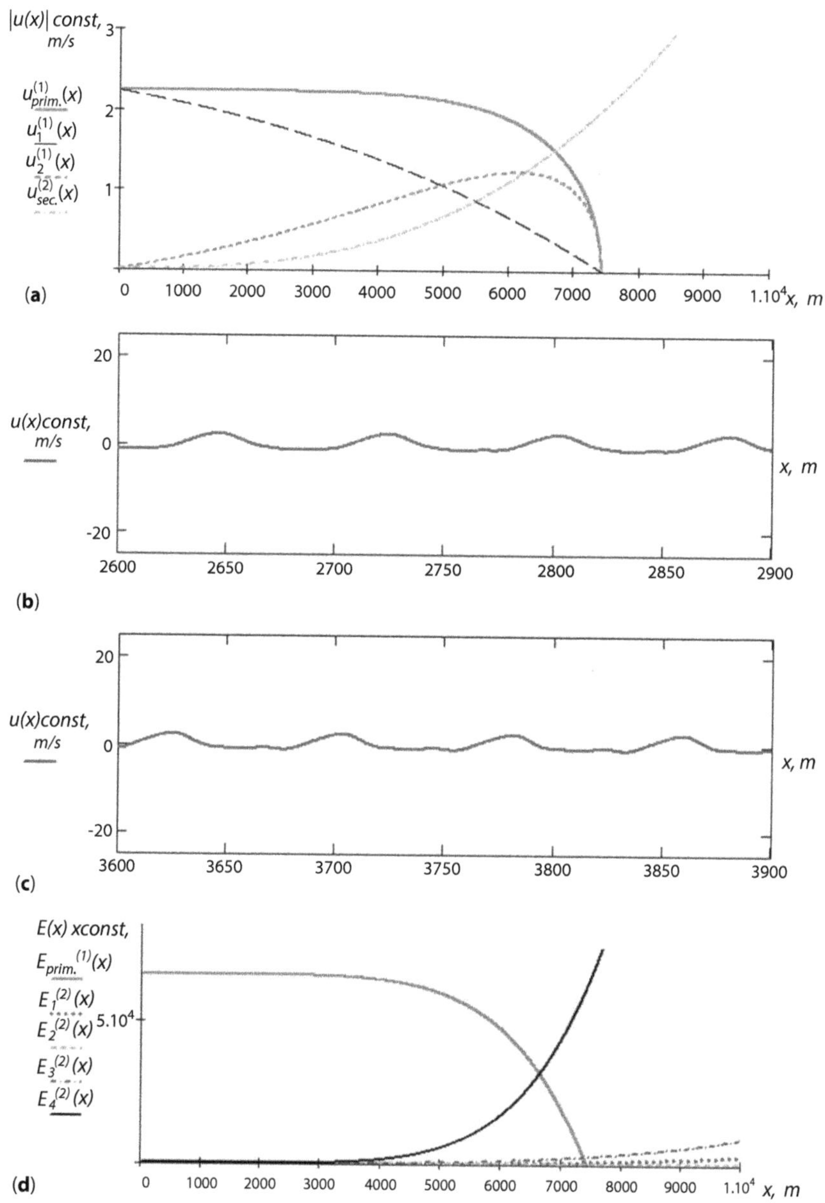

Figure 3.4.4 Main characteristics of the surface gravity waves with parameters: $f = 0.09$ Hz; $\lambda = 77.8$ m; $c = 7$ m/s; H = 5 m; $2a/\lambda = 0.015$; $kH = 0.4$; $\varepsilon = 0.118$; $\gamma = 0.004$.

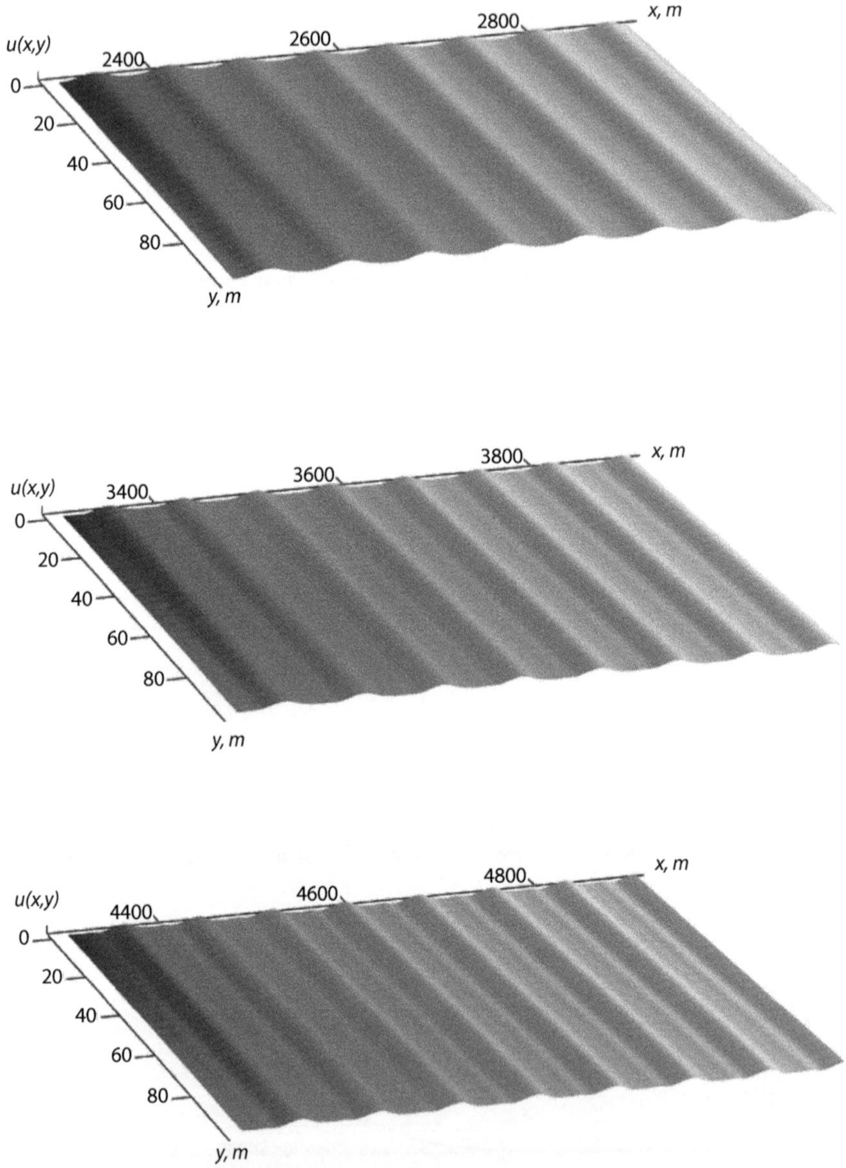

Figure 3.4.5 Spatial model of the propagation of a surface gravity wave with parameters: $f = 0.09$ Hz; $\lambda = 77.8$ m; H = 5 m; $2a/\lambda = 0.014$; $\varepsilon = 0.107$; $\gamma = 0.004$.

speed under the crests is higher than under the troughs. These features are explained by the Stokes theory by the appearance of high-frequency harmonics of the fundamental wave. Such profiles acquire surface waves for our consideration at the initial propagation sites, which can be observed in

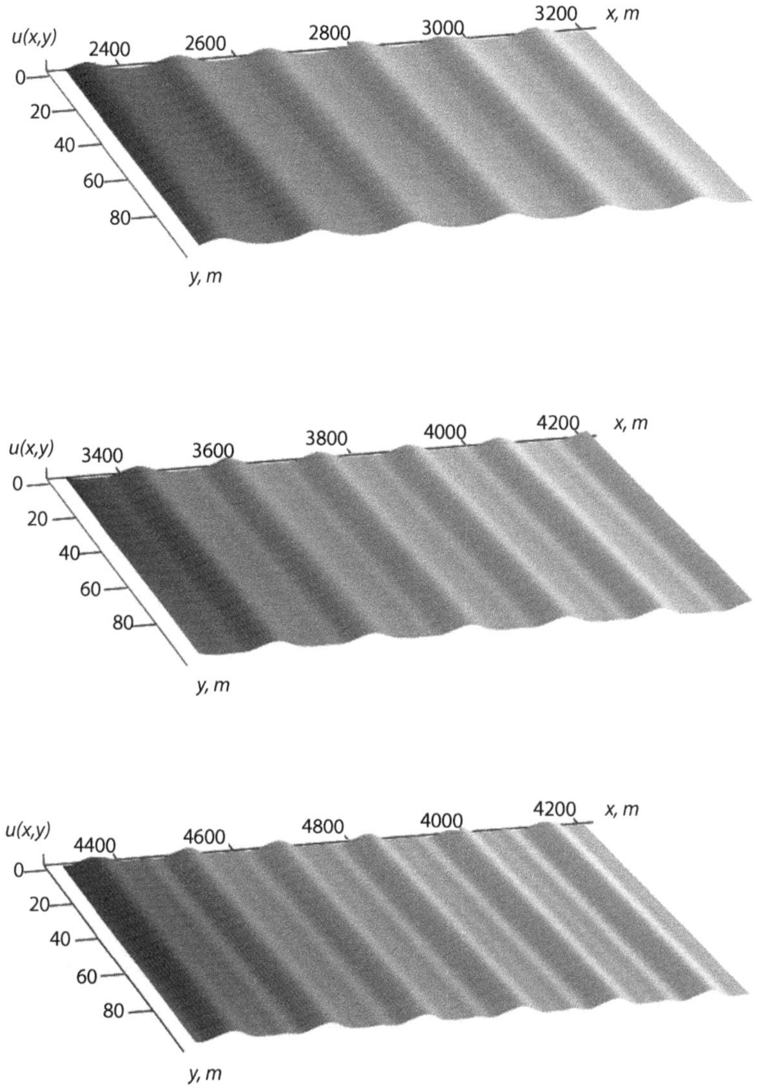

Figure 3.4.6 Spatial model of the propagation of a surface gravity wave with parameters: $f = 0.045$ Hz; $\lambda = 155.6$ m; H = 5 m; $2a/\lambda = 0.0088$; $\varepsilon = 0.136$; $\gamma = 0.001$.

Figure 3.4.1b, 3.4.3b, 3.4.4b. At this stage, the generation of higher harmonics leads to a sharpening of the crests, while the crests themselves remain symmetrical with respect to the vertical axis.

In addition to the described type of asymmetry, another type of asymmetry in the profile of real surface waves is still observed. It is expressed in the fact that the front slope of the transforming wave becomes steeper than the rear one [Flick, 1981], [Peregrine, 1967], [Zaharov, 1998]. This difference in the steepness of the slopes increases with the approach to the shore. This type of asymmetry in the profile of real surface waves is not described by the Stokes theory, nor by most other theories, although an equation with an integral term on the basis of the Korteweg-de Vries equation was proposed by J. Whitham to describe the process of waves from the high seas to the shore. At the stage of wave breaking, the action of the integral term becomes insignificant and the solutions of the Whitham equation behave similarly to solutions of the simplest equations with a nonlinearity (analogous to (3.3.1)), breaking and collapsed over a finite time [Whitham, 1977], [Gabov, 1988].

For our consideration, the surface wave profiles in the third propagation section acquire a second asymmetry, Figure 3.4.1c, 3.4.3c, 3.4.4c. Thus, steepening becomes more pronounced with increasing wavelength, i.e., with increasing nonlinearity coefficient.

One explanation for the steepening of the front slope of the crests is that the higher harmonics gradually shift in phase relative to the main wave [Flick, 1981], [Lonquet-Higgins, 1993], [Zacharov, 1998], [Leontiev, 2001], [Kuznetsov, 2005]. When the phase shift reaches a value of $\pi/2$, the leading slope becomes almost vertical, causing the wave to collapse. In the final expression obtained for the horizontal particle velocity (3.3.16), the amplitudes of the even harmonics are imaginary. As the amplitude of the secondary waves increases, this imagination begins to be felt stronger than at the beginning, since it leads to a phase shift relative to the amplitude of the fundamental wave. It should be noted that in the Stokes theory the phase difference between the fundamental wave and the higher harmonics is absent, i.e., it cannot describe these processes (expression 2.2.9).

To determine the effect of even harmonics, and especially the fourth harmonic, on the surface gravity wave profile in Figure 3.4.7, graphs of the horizontal velocity $u(x,t)$ of the medium particles for the same section of the propagation path are presented. The special influence of the fourth harmonic is due to its fastest growth, hence, the highest energy intensity among the secondary waves.

Figure 3.4.7a shows the surface wave profile calculated according to the expression (3.3.16), when the fourth harmonic in phase lags behind the main wave by $\pi/2$. Figure 3.4.7b shows the wave profile for the case when the imaginary amplitude of the fourth harmonic is replaced by the reality, i.e., phase shift between the harmonic and the fundamental wave is absent. In this case, the amplitude of the fourth harmonic has a negative sign; therefore, it is subtracted from the amplitude of the fundamental wave. This leads to the symmetry of the wave crest relative to the vertical axis.

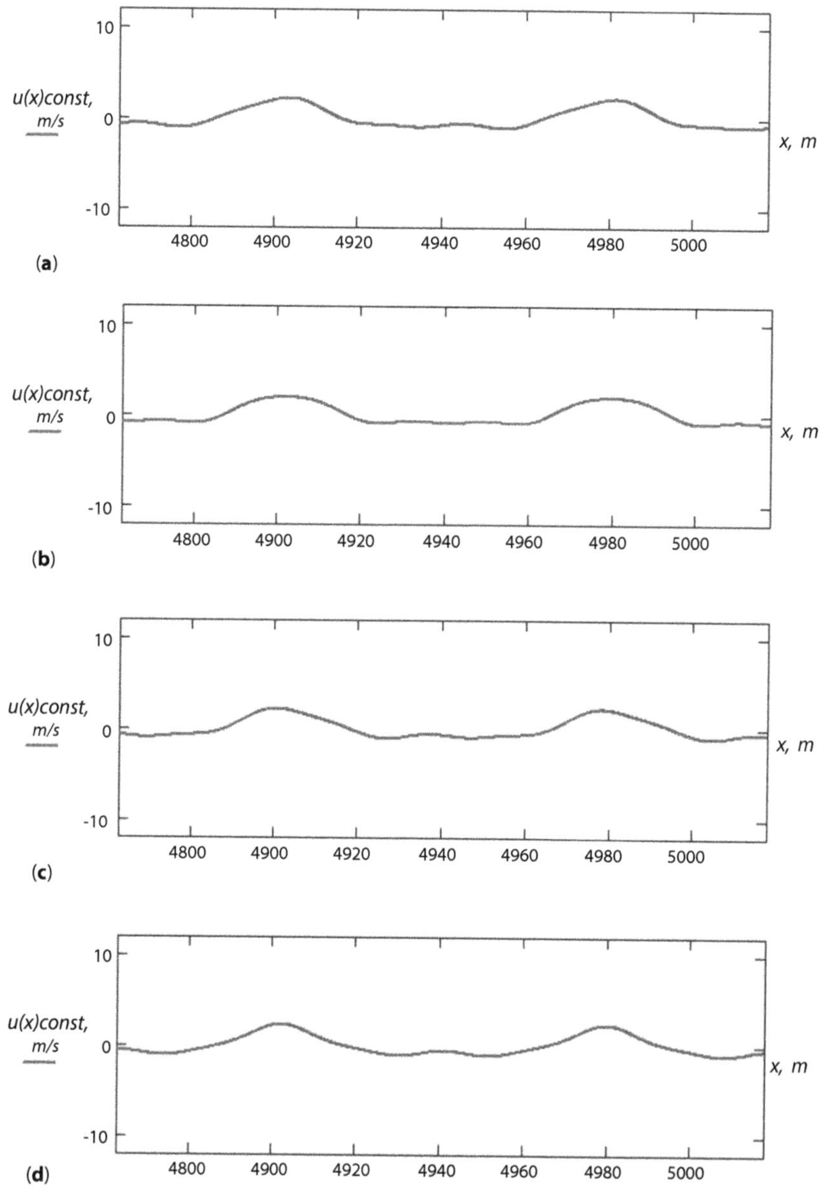

Figure 3.4.7 The influence of the fourth harmonic on the profile of the surface gravitational wave: (a) lags behind the main wave by $\pi/2$; (b) in-phase subtracted; (c) ahead of the main wave by $\pi/2$; (d) – is in-phase summed. Wave parameters: f = 0.09 Hz; $\lambda = 77.8$ m; $c = 7$ m/s; H = 5 m; $2a/\lambda = 0.014$; $kH = 0.4$; $\varepsilon = 0.107$.

In expression (3.3.16), the amplitude of only the fourth harmonic has a negative sign. Replacement of a negative sign with a positive sign leads to a change in the steepening of the leading edge of the crest of the wave to the trailing edge. This case is shown in Figure 3.4.7c, which resembles the profile of a wave turned back, the wave propagates in the opposite direction.

In this case, the fourth harmonic is ahead of the main wave by $\pi/2$ in phase. If in the expression (3.3.16) we simultaneously replace the negative sign by a positive sign, also the imaginary of the fourth harmonic by the reality, then the wave profile takes the form as in Figure 3.4.7e. In this case, the amplitude of the fourth harmonic is summed with the amplitude of the main wave in phase, amplifying the sharpening of the wave crest. This profile resembles the Stokes waves of finite amplitude in deep water (or trochoidal wave).

Attention should also be paid to the appearance of a local elevation in the trough between the crests. This elevation is also associated with the fourth harmonic, and it changes its localization depending on the phase. Experimental observations with the emergence of an elevation between the main wave crests in the coastal zone are presented in [Goda, 1998], [Kawasaki, 1999], [Kuznetsov, 2002].

Another explanation for the steepness of the foreland is due to the difference in the speeds of the crests and troughs of the waves in shallow water. Under the crests, the total depth is equal $H+a/2$, but under the troughs $H-a/2$. Under the crests, the depth is greater and, consequently, the velocity of the wave is higher than in the troughs [Lemb, 1947]. Over time, the crests will crawl to the hollows, as a result of which the steepness of the front slope will increase. This explanation reveals the initial mechanism for the onset of nonlinearity in shallow water, and the description of the physics of the process will lead to the wave variant of the description.

Consequently, the proposed method makes it possible to describe both the process of initial sharpening of the crests and the subsequent steepening of the leading edge. The increase in the velocity of the particles on the crest leads to further steepening of the leading edge. When the velocity of the particles becomes comparable with the phase velocity of the wave, it becomes unstable and begins to collapse.

In shallow water with a flat bottom, caving can occur as a diving or sliding breaker. The diving surf is accompanied by the ejection of the jet from the crests, followed by the capture of air (Figure 3.4.8).

With a sliding breaker, the upper part of the ridge flows downward in the form of a roller, creating considerable turbulence. If the collapse occurs away from the shore, the surface wave gets rid of the accelerated particles of the medium on the steep leading edge of the crest, and this leads to a

Figure 3.4.8 The steepening of the leading edge of the wave crest and the formation of a diving breaker (ejection of a jet from the crest with the entrapment of air) (*Durso village*).

decrease in the wave height. In the future, it will finally collapse on the shore, like all other surface nonlinear waves with a steep leading edge. Within the framework of the linear approximation for small amplitude values, wave dissipation can occur without its collapse (Figure 2.6.3).

The collapse of the wave on the shoreline leads to the phenomenon of the surf (Figure 3.4.9), when the potential energy of the wave goes into a pleasant surf noise, the appearance of turbulence with foam, the movement of the coastal suspended mass.

3.5 Modeling the Refraction of Nonlinear Surface Gravity Waves

Let us consider the wave processes occurring when nonlinear gravity surface waves approach the shore in shallow water conditions. We simulate the coastal slopes of different steepness in the conditions of the bay. The bottom lines are shown in Figure 3.5.1 (depth scale is increased). They are modeled on the basis of the graphs of the functions: sine – $s(x) = x\sin x$ and Bessel of the fifth order – $s(x) = x\,\mathrm{J}(5, x)$ in the range from 0 to 10 km along the x axis. The depth decreases from $H = 5$ m to zero, the slope steepness does not exceed $0{,}01^0$. The bottom line based on the sine function is flat along the entire length of the bottom, and on the basis of the Bessel function steeper near the shore.

Figure 3.4.9 The breaking of the crest of a wave of swell on the shore (*Durso village*).

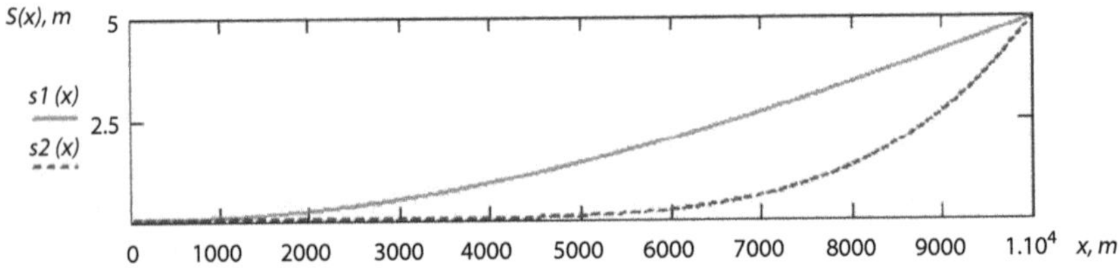

Figure 3.5.1 Bottom lines according to analytic expressions: $s1(x) = 0.0006x \times \sin(0.0001x)$; $s2(x) = 2x \times J(5, 0.0001x)$.

As was already noted, the depth gradually decreases in the process of surface-wave propagation. However, the formula (2.3.17) for the surface gravity wave propagation velocity is derived for the case when the depth of shallow water is constant [Lamb, 1947]

$$c = \sqrt{gH}$$

However, if the depth changes within the length of a single wave only by a fraction, formula (2.3.17) is acceptable for slowly changing depths [Gabov, 1988], [Shuleikin, 1968]. In our case, the average change in depth for the highest-frequency wave (satisfying the condition of shallow water $H/\lambda << 1$) with the wavelength $\lambda = 30$ m is less than 0.01λ.

When a gravity wave propagates in a bay with such a bottom line, its wavelength gradually decreases, and in the end, it breaks (on the coast or before reaching it). The wavelength decrease follows from the expression

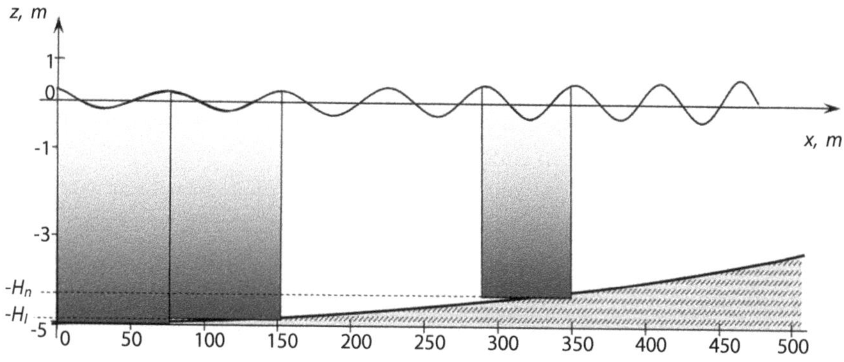

Figure 3.5.2 Geometry of the problem related to the propagation of a surface gravity wave along a sloping bottom in conditions of a bay.

for the wave number (2.3.17) then, taking into account the change in depth, we obtain

$$k(x)=\sqrt{\frac{\omega^2}{g(H-s(x))}},\lambda(x)=\frac{2\pi}{k(x)}. \tag{3.5.1}$$

The dependence of the wave number $k(x)$ and the velocity $c(x)$ on the depth leads to a dependence on the distance (due to a decrease in depth) and other basic parameters of the surface wave: wavelength $-\lambda(x)$; amplitudes $-a(x)$; nonlinear parameter $-\varepsilon(x)$; dispersion $-\gamma(x)$.

It should be noted that, with an increase in the length of a surface wave as it approaches the coast, the wave front also becomes distorted. However, the law of conservation of energy must be observed. To analyze the main parameters of the surface gravity wave, we use the expression (2.3.23) for its total energy

$$E=\frac{\rho_0 g}{4}\lambda a^2$$

A decrease in the wavelength leads to an increase in the displacement amplitude, because the law of energy conservation must be observed for the wave process (the expression (3.4.2)).

The geometry characterizing the problem of surface wave propagation toward the coast in conditions of a bay is illustrated by Figure 3.5.2. Within each wavelength, the bay volume is divided into vertical columns of liquid.

Each column is restricted from below by the plane $z = -H_n$, and from above, by the free surface. In the direction of wave travel, the column is restricted by the planes $x = x_n$ and $x_{n+1} = x_n + \lambda$, where as in the crosswise direction, the planes are spaced by a unit of length. The law of energy conservation is observed within each column of liquid.

In order to trace the distortion of the profile of a nonlinear surface gravity wave, we use the expression for the horizontal velocity of particles of the medium (3.3.16) which was obtained by the method of successive approximations:

$$\begin{aligned} u_{\text{sec.}}(x,t) &= \varepsilon u^{(1)}_{prim.}(x,t) + \varepsilon^2 u^{(2)}_{\text{sec.}}(x,t) \\ &= \left[\varepsilon^2 U_0 \exp(ikx) - ikt\varepsilon^3 U_0^2 \exp(2ikx) \right] \\ &+ \left[k^2 t^2 \varepsilon^5 U_0^3 \exp(ikx) + ikt\varepsilon^4 U_0^2 \exp(2ikx) + \right. \\ &\left. + 3k^2 t^2 \varepsilon^5 U_0^3 \exp(3ikx) - 2ik^3 t^3 \varepsilon^6 U_0^4 \exp(4ikx) \right] + (c.c.) \end{aligned}$$

The distributions of the oscillatory velocities of the primary $u^{(1)}_{prim.}(x,t)$ and secondary waves $u^{(2)}_{\text{sec.}}(x,t)$, when the gravity wave propagates along the bay with the bottom line based on the Bessel function, are shown in Figure 3.5.3, and in Figure 3.5.4. – with a bottom line based on the sine function. The values of the initial parameters of this surface wave are analogous to the parameters of the surface wave, presented in Figure 3.4.3: $\lambda =$ 77.8 m; $2a/\lambda = 0.014$.

Since in this case the depth of the bay decreases, hence, the process of breaking of the surface wave is accelerated because of the increase in nonlinearity. This wave, when spreading in a bay with a flat bottom, came into instability at a distance of – $x^{(2)}_{char.} \approx 10$ km.

It also comes to instability at a distance of $x^{(2)}_{char.} \approx 7.3$ km for slowly rising at the beginning of the bottom (based on the Bessel function), and our model is correct to a distance of $x \approx 5$ km (Figure 3.5.3). The instability for a uniformly rising bottom (based on the sine function) occurs at a distance of – $x^{(2)}_{char.} \approx 5$ km, and the correct model is up to $x \approx 3.8$ km (Figure 3.5.4).

In this case, up to a distance $x \approx 7$ km, the depth based on the Bessel function is an order of magnitude smaller than the depth based on the sine function, so the surface wave for such a bottom is more stable. The decrease in the instability distance is associated with an increase in the magnitude of the displacement amplitude (because of the decrease in the wavelength).

In comparison with a flat bottom, the amplitude of primary waves with a distance more slowly decreases. It should be noted that, in passing to each

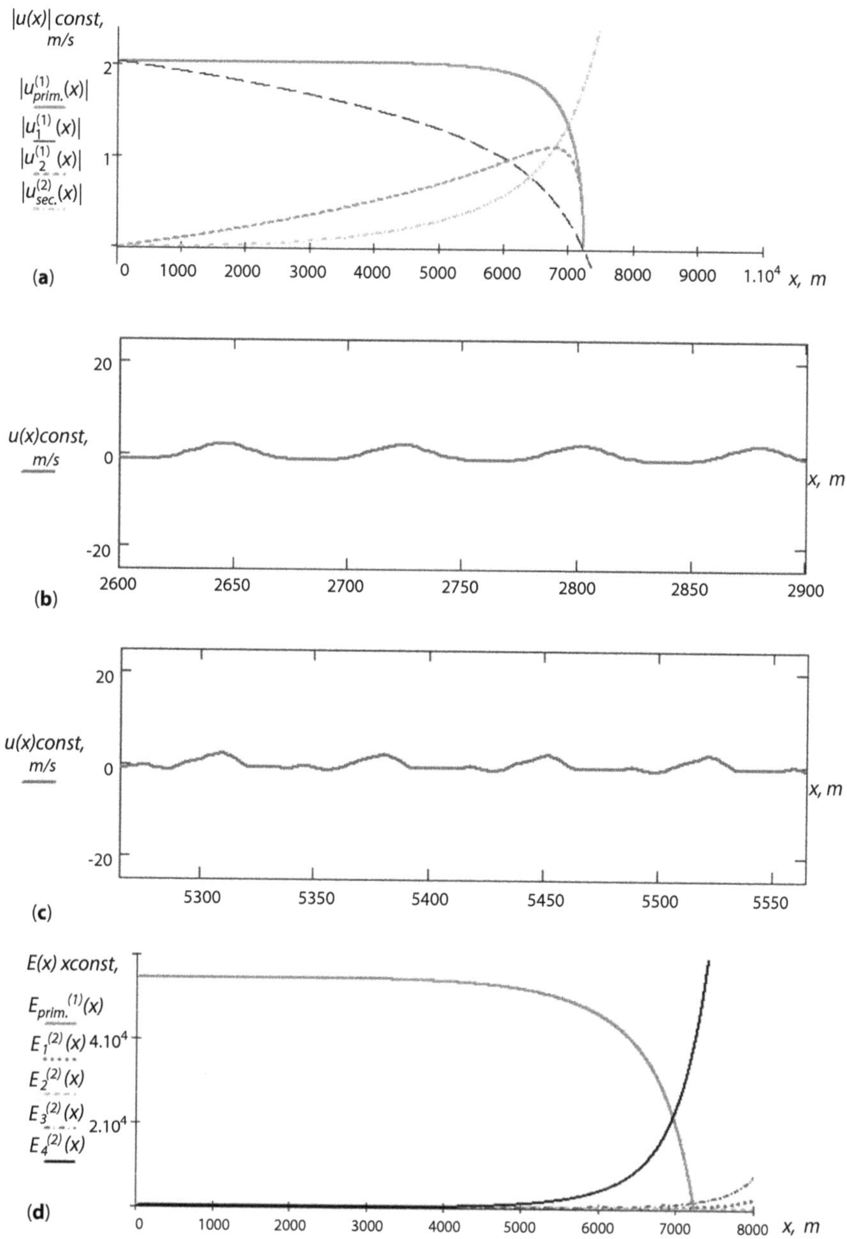

Figure 3.5.3 The main characteristics of a surface gravity wave with initial parameters: $f = 0.09$ Hz; $\lambda = 77.8$ m; $c = 7$ m/s; $2a/\lambda = 0.014$; $kH = 0.4$; $\varepsilon = 0.107$; $\gamma = 0.004$, bottom line based on the Bessel function.

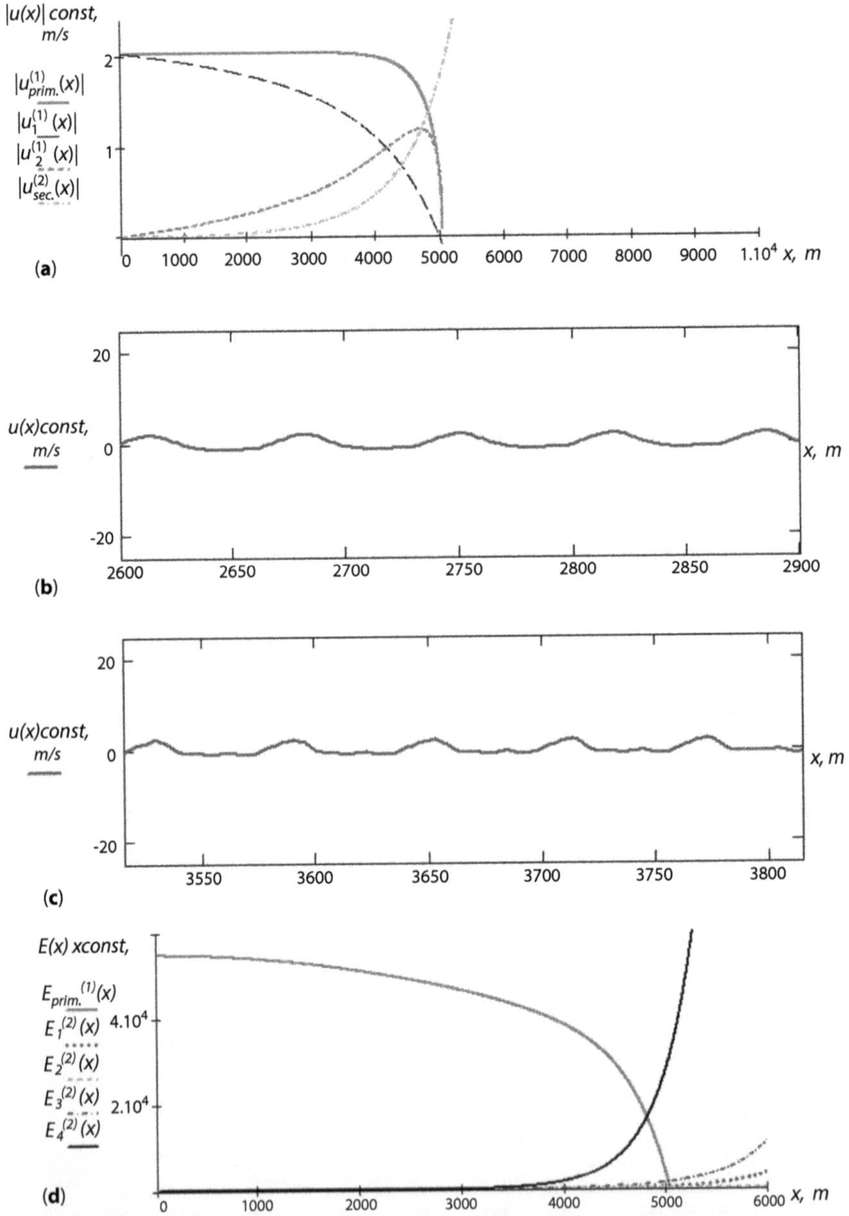

Figure 3.5.4 The main characteristics of a surface gravity wave with initial parameters: $f = 0.09$ Hz; $\lambda = 77.8$ m; $c = 7$ m/s; $2a/\lambda = 0.014$; $kH = 0.4$; $\varepsilon = 0.107$; $\gamma = 0.004$, bottom line based on the sine function.

new column of liquid (Figure 3.5.2), the displacement amplitude increases. First, the wavelength is determined by formula (3.5.1) for each new column, whereupon, taking into account the constant energy, new values are found for the amplitude of the free-surface displacement, nonlinearity parameter, and oscillatory velocity of particles. The nonlinearity parameter $\varepsilon_n(x) = a_n(x)/H_n(x)$ also increases in this case, because the depth decreases with distance.

The distortion of the profile of the nonlinear surface gravity wave (3.5.3b, c, and 3.5.4b, c) occurs more rapidly, and with a larger amplitude. Moreover, from the comparison of Figures 3.5.3b and 3.5.3c (also Figures 3.5.4b and 3.5.4c), one can notice a gradual decrease in the wavelength (wave compression), which by itself further intensifies the steepness of the wave.

It is also seen from Figures 3.5.3e and 3.5.4e that the energy of the primary waves (without dissipation) to the onset of instability is completely transformed into the energy of the secondary waves.

For three-dimensional modeling of the refraction of a nonlinear surface gravity wave on coastal formations, we will create a three-dimensional model of the cape, bay and flat coastline.

Figure 3.5.5 shows a model of a three-dimensional cape with dimensions of 10 km × 10 km × 25 m (scale on the z-axis is increased). It is described by an analytic expression

$$z(x, y) = a\sqrt{xy}\sin(bxy). \quad (3.5.2)$$

Figure 3.5.6 shows the spatial model of the bay according to the analytic expression

$$z(x, y) = ay^2\cos(bx) + (cy)^2. \quad (3.5.3)$$

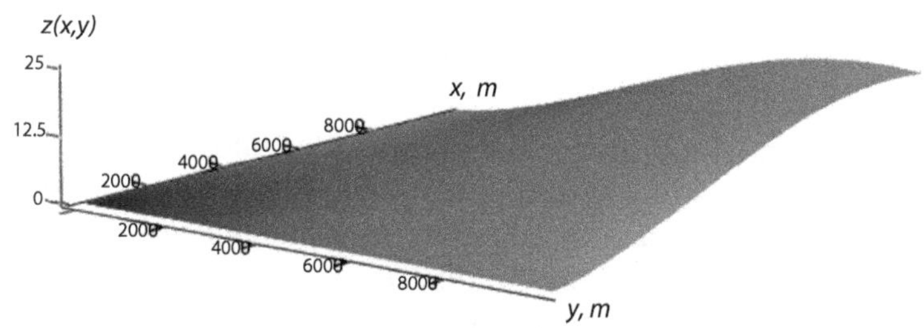

Figure 3.5.5 Spatial model of the cape (10 km × 10 km × 25 m) by analytic expression $z(x,y) = 0.0024(xy)^{1/2} \times \sin(2 \times 10^{-8}xy)$.

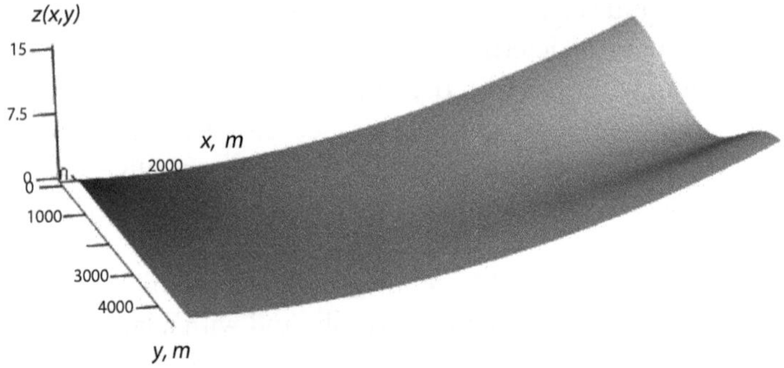

Figure 3.5.6 Spatial model of the bay (10 km × 10 km × 15 m) by analytic expression $z(x,y) = 3 \times 10^{-8} y^2 \cos(0{,}0014x) + (3.4 \times 10^{-4} y)^2$.

The spatial model of a straight coastal protrusion is described by an analytic expression

$$z(x,y) = 0.0006y \times \sin(0.0001y). \tag{3.5.4}$$

To simulate the process of refraction of surface gravity waves on these shore formations, let us use the expression for the wave number in the two-dimensional case (based on the expression (3.5.1)):

$$k(x,y) = \sqrt{\frac{\omega^2}{g(H - z(x,y))}}, \tag{3.5.5}$$

where $z(x,y)$ – analytic expression of the relevant shore formation.

Such coastal formations exist in the Bay of Taganrog in the areas of the Beglitskaya and Aleksandrovskaya spits.

Figure 3.5.7 shows a three-dimensional model of refraction of a nonlinear surface gravity wave on the cape in the range from 0 to 4.5 km along the x and y axes. The gravity wave has the following initial parameters: f = 0.045 Hz; λ = 155.6 m; H = 5 m; $2a/\lambda$ = 0.0076; approach angle of the wave is $\theta = 0^0$ relative to the axis x.

For calculations, the surface of the bay is divided into a grid with a pitch of 8.5 m along the x and y axes, the step value is one order of magnitude smaller than the sampling interval of the corresponding wave (at least 10 points per wavelength [Gonorovsky, 1957]).

The wave vector is directed along the x axis. Spatially three-dimensional models are created by the method of plane-parallel transfer of the initial velocity field along the y axis [Lavrent'ev, 1973].

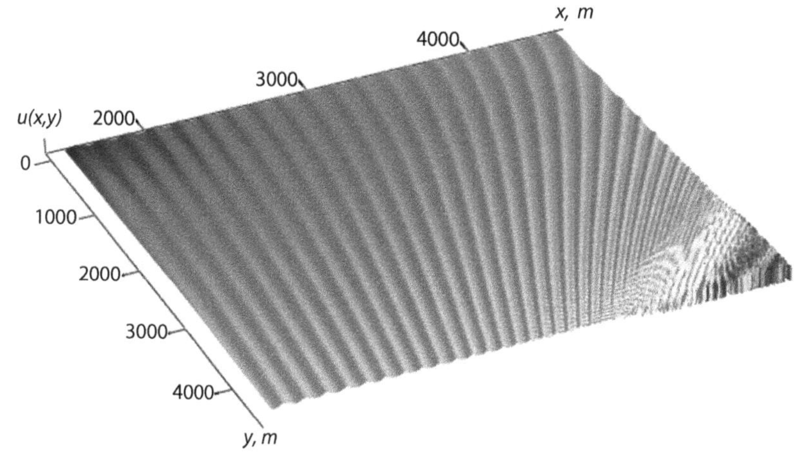

Figure 3.5.7 Three-dimensional model of refraction to the cape of a nonlinear surface gravity wave with initial parameters: $f = 0.045$ Hz; $\lambda = 155.6$ m; $H = 5$ m; $2a/\lambda = 0.0076$; $\theta = 0^0$.

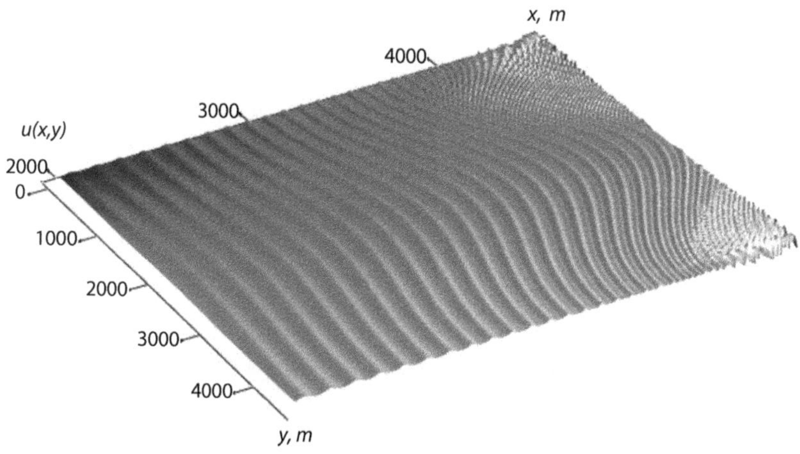

Figure 3.5.8 Three-dimensional model of refraction in the bay of a nonlinear surface gravity wave with initial parameters: $f = 0.045$ Hz; $\lambda = 155.6$ m; $H = 5$ m; $2a/\lambda = 0.0076$; $\theta = 0^0$.

As the front of the nonlinear surface wave approaches the cape, it starts to assume a curvilinear shape and tends to become normal to the coast line. As for the direction of the x axis, the wave profile, along with a decrease in the wavelength, undergoes a distortion, and its steepness progressively increases.

Figure 3.5.8 shows a three-dimensional model of refraction in the bay of a nonlinear surface gravity wave: angle $\theta = 0^0$, the extent of the bay along the x axis is from 2 km to 4.5 km, along the y axis is from 0 to 4.5 km.

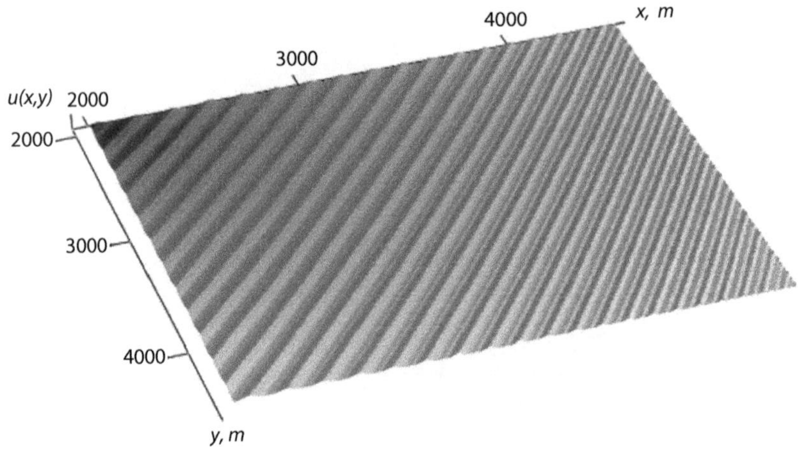

Figure 3.5.9 Three-dimensional model of refraction on the flat coastline of a nonlinear surface gravity wave with initial parameters: $f = 0.045$ Hz; $\lambda = 155.6$ m; $H = 5$ m; $2a/\lambda = 0.0076$; $\theta = 45^0$.

Figure 3.5.10 The fragment of the three-dimensional refraction model on the cape of a nonlinear surface gravity wave with the approach angle $\theta = 45^0$.

Figure 3.5.9 shows a three-dimensional model of refraction on the flat side of a nonlinear surface gravity wave: the approach angle $\theta = 45^0$ relative to the axis x, the extent of the bay along the x axis and the y axis is from 2 km to 4.5 km.

Figure 3.5.10 shows an enlarged fragment of a three-dimensional refraction model on the cape of a nonlinear surface gravity wave with an approach angle $\theta = 45^0$ relative to the axis x.

3.6 Modeling of Propagation and Refraction of Nonlinear Surface Gravity Waves Under Shallow Water Conditions with Account of Dispersion

Let us consider the problems of the propagation of nonlinear surface gravity waves in shallow water bays with account of dispersion. Under the shallow-water condition nonlinearity – ε and dispersion – γ is determined with expression 3.2.1

$$\varepsilon = \frac{\vartheta_x}{c} = \frac{a}{H}, \gamma = \left(\frac{H}{\lambda}\right)^2,$$

where ϑ_x is the horizontal velocity of particles, c is the propagation velocity of gravity waves, a is the amplitude of the vertical shift of the free surface, H is the fluid depth, λ is wavelength.

In the absence of dispersion by the method of successive approximations, the expression (3.3.16) was obtained by solving the shallow water equation with a quadratic nonlinearity. Ignoring the dispersion was connected with its weakening under shallow-water conditions. Though, even small dispersion under shallow-water conditions can prevent wave breaking. And the small dispersion balance with small nonlinearity leads to stationary cnoidal waves (3.2.18).

The dispersion influence on the propagation process will be taken into account in the original, through the well-known dispersion law for circular frequency of surface gravity waves:

$$\omega = \sqrt{gk \cdot \mathrm{th}\, kH}$$

phase velocity of gravity waves is determined from the expression:

$$c_{ph} = \frac{\omega}{k} = A_0 \cdot \sqrt{\frac{\mathrm{th}\, kH}{kH}}$$

where $c_0 = \sqrt{gH}$ – is propagation velocity of surface waves in shallow-water approximation. From expression (2.3.9) it follows that the propagation velocity for appearing secondary waves with the wave shortening will be reduced.

In the first approximation we consider that synchronism condition is fulfilled for primary interacting waves: fundamental and second harmonic, i.e. $\omega_2 = 2\omega$ and they propagate with common velocity c_0 the solution is given by

$$u^{(1)}_{prim.}(x,t) = \varepsilon u^{(1)}(x,t) + \varepsilon^2 u^{(2)}(x,t)$$
$$= \varepsilon U_0 \exp\left[ikx\right] - ikt\varepsilon^2 U_0^2 \exp\left[i2kx\right] + c.c.$$

As a consequence of interaction between fundamental and second harmonics secondary waves will occur, which lead to steepening of wave front. When solving the nonlinear equation (3.3.1) by the method of successive approximations, the squaring and differentiation of the solution in the first approximation (expression (3.3.10)) is carried out

$$u^{(2)}_{sec.}(x,t) = \left[\left(t\varepsilon^2 U_0^2/2\right)\cdot ik_2 \exp\left[ik_2 x\right] + \left(-k^2 t^3 \varepsilon^4 U_0^4/2\right)\cdot ik_4 \exp\left[ik_4 x\right]\right.$$
$$\left. + \left(-ikt^2\varepsilon^3 U_0^3\right)\cdot ik_3 \exp\left[ik_3 x\right] + \left(-ikt^2 \varepsilon^3 U_0^3\right)\cdot ik_1 \exp\left[ik_1 x\right]\right]$$
$$+ (c.c.) \qquad (3.6.1)$$

For taking the dispersion into account in this expression, wave numbers of secondary waves are specified by applicable indexes:

$$k_n = \frac{\omega_n}{c_n}, \quad \omega_n = n\omega, \quad n = 1, 2, 3, 4 \text{ – harmonic number,}$$

thereat, every wave will propagate with its velocity:

$$c_n = A_0 \cdot \sqrt{\frac{\text{th}(nkH)}{(nkH)}} \qquad (3.6.2)$$

In view of expression (3.6.1) the solution in two approximations for horizontal velocity of medium particles will be given by

$$u_{sec.}(x,t) = \varepsilon u^{(1)}_{prim.}(x,t) + \varepsilon^2 u^{(2)}_{sec.}(x,t)$$
$$= \left[\varepsilon^2 U_0 \exp(ikx) - ikt\varepsilon^3 U_0^2 \exp(2ikx)\right]$$
$$+ \left[\left(-ikt^2 \varepsilon^5 U_0^3\right)\cdot ik_1 \exp\left[ik_1 x\right] + \left(t\varepsilon^4 U_0^2/2\right)\cdot ik_2 \exp\left[ik_2 x\right]\right.$$
$$\left. + (-ikt^2 \varepsilon^5 U_0^3)\cdot ik_3 \exp[ik_3 x] + (-k^2 t^3 \varepsilon^6 U_0^4/2)\cdot ik_4 \exp[ik_4 x]\right]$$
$$+ (c.c.) \qquad (3.6.3)$$

Amplitude velocity components of secondary waves have different power law from time (second term of expression (3.6.3)):

$$u_n^{(2)}(x,t) = U_{n0}(t)\exp(ik_n x). \qquad (3.6.4)$$

Phasic part of velocity has a wave number, which will depend on velocity of every harmonic because of expression (3.6.2). Therefore, for high harmonics propagation velocity will be reduced; it's the result of dispersion. Therefore, secondary waves, appearing as a result of nonlinear effects, will not lead to the steepening of wave front because of discordance.

Amplitude components also depend on harmonic propagation velocity through temporary element and wave number. With account of inverse proportionality and propagation velocity, harmonic amplitude will increase rapidly with time than when the dispersion is present.

It is necessary to note that contrary to cnoidal waves, in our case the dispersion effect on the profile of surface wave could be observed in dynamics with its propagation on shallow waters.

To check the effect of dispersion on nonlinear wave processes we will use hydrologic conditions of the Bay of Taganrog of the Azov Sea. The mean depth of the bay does not exeed 5 m, therefore, the shallow-water conditions will be satisfied for gravity waves with their lengths longer 30 m, the bay floor is flat, surface tension is absent, the effect of wind is not taken into account.

Figure 3.6.1a gives graphs of behavior from distance amplitudes of primary waves $u_{prim.}^{(1)}(x,t)$ and secondary waves $u_{\text{sec.}}^{(2)}(x,t)$ velocities, as well as fundamental wave $u_1^{(1)}(x,t)$ and second harmonic $u_2^{(1)}(x,t)$.

For the surface wave with initial parameters: length λ = 77.8 m; initial steepness $2a/\lambda$ = 0.014, the distance of instability is $x_{char.}^{(2)} \approx 6600$ m. The values of the initial parameters of this surface wave are similar to the parameters of the surface wave presented in Figure 3.4.3.

For a surface wave with initial values: λ = 77.8 m; $2a/\lambda$ = 0.014, disregarding the variance, the instability occurred at a distance of $x_{char.}^{(2)} \approx 10$ km, taking this dispersion into account, this distance decreases to $x_{char.}^{(2)} \approx$ 6600 m.

The decrease of instability distance is connected with the rapid growth of secondary wave amplitude because of its inverse dependence on the propagation velocity. With account of dispersion the propagation velocity of secondary waves, because of expression (3.6.2), takes the following values: $c_1 = 6.8$ m/s; $c_2 = 6.4$ m/s; $c_3 = 5.8$ m/s; $c_4 = 5.3$ m/s. The propagation velocity within the shallow-water model is $c_0 = 7$ m/s, thus velocity change for the fourth harmonic will be $\Delta c_4 = 1.7$ m/s, i.e. it will be 25% less.

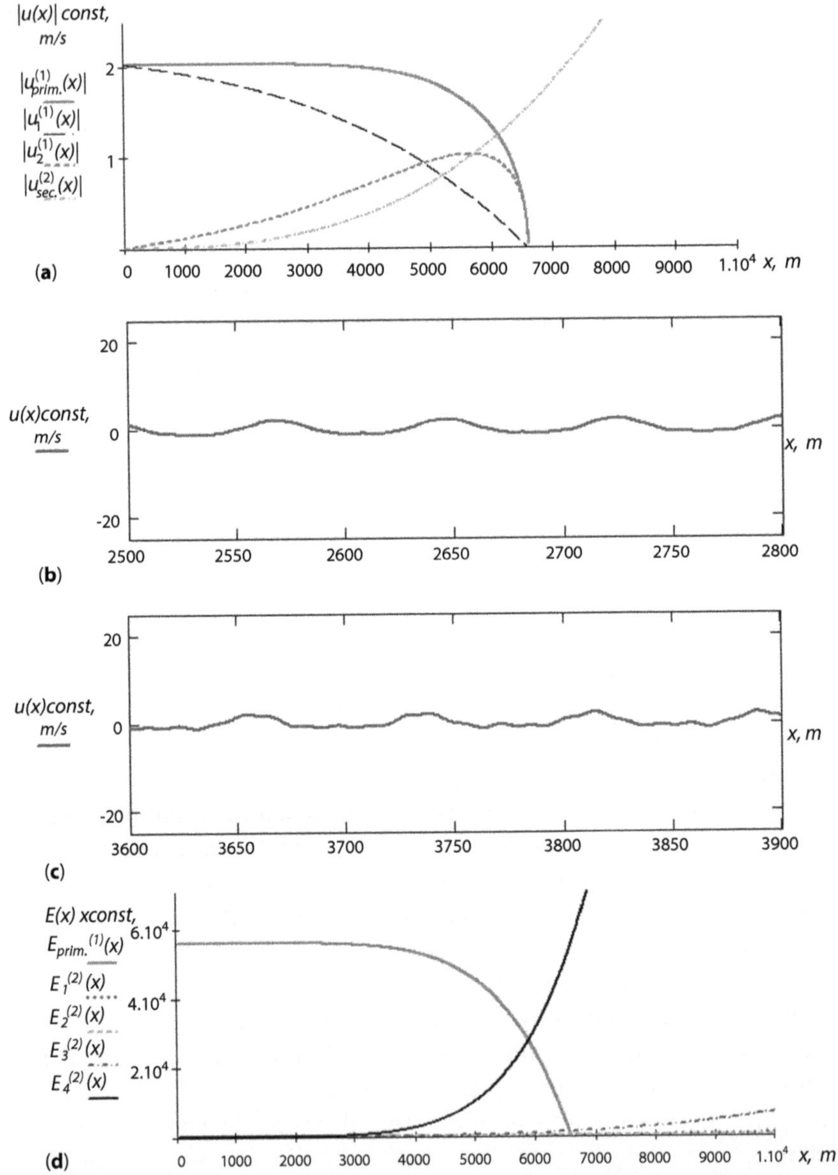

Figure 3.6.1 The main characteristics of a surface gravitational wave with allowance for variance: $f = 0.09$ Hz; $\lambda = 77.8$ m; $2a/\lambda = 0.014$; $kH = 0.4$; $\varepsilon = 0.107$; $\gamma = 0.004$.

It follows from Figure 3.6.1a, the summary amplitude of secondary waves increases with the decrease of primary waves amplitude. The amplitude of the second harmonic slowly increases, and by the moment of instability, rapidly decreases, as the energy fully transfers from primary (fundamental wave and second harmonic) to secondary waves.

Let us consider variations in the profile of the initial gravity wave over the distance $x \approx 4000$ m, where the initial applying condition of method of successive approximation is performed $u^{(n+1)} << u^{(n)}$, i.e., the summary amplitude of secondary waves is considerably less than the amplitude of primary waves. Thus, this distance is still considered to be correct.

In Figure 3.6.1b and 3.6.1c give the path-length dependence of variations in the horizontal velocity $u(x,t)$ on the bases of the expression (3.6.3). Over the travel time, the wave with an initially cosine profile is gradually distorted, the crests steepen, and the wave troughs flatten (Figure 3.6.1b). It is caused by nonlinear processes leading to the higher harmonics. Further, the higher harmonics are affected by dispersion. Shallow-water conditions are not fulfilled for the third and fourth harmonics, and they start propagating with less velocity and it leads to the loss of synchronism.

As a result, if nonlinearity doesn't bring the wave to breaking (to critical values of cutting-point angle), the dispersion will bring the initial packet of waves to breaking, as it's given in Figure 3.6.1c. Figure 3.6.1d gives energy characteristics of primary and secondary waves, where we can observe energy conservation principle in wave process.

Further, we discuss the dispersion effect on the propagation of surface gravity waves under shallow-water conditions with various wave parameters. Figure 3.6.2, 3.6.3 and 3.6.4 give profiles of surface gravity waves depending on travel distance for three different waves.

For comparison Figure 3.6.2a, 3.6.3a and 3.6.4a give profiles of surface waves without account of dispersion (based on expression 3.3.16), and Figure 3.6.2b, 3.6.3b and 3.6.4b give it with account of dispersion (based on the expression 3.6.3). Values of travel distances in the Figure s were chosen on the principle of equal energetic levels of secondary waves.

For surface wave λ = 35 m (Figure 3.6.2) the propagation velocity of secondary waves takes the following values: c_1 = 6.3 m/s; c_2 = 5.1 m/s; c_3 = 4.3 m/s; c_4 = 3.7 m/s. Difference of the propagation velocity for the fourth harmonic will be Δc_4 = 3,3 m/s, i.e., it is 47% less.

For surface wave λ = 155.6 m (Figure 3.6.4) the propagation velocity of secondary waves takes the following values: c_1 = 6.9 m/s; c_2 = 6.2 m/s; c_3 = 6.6 m/s; c_4 = 6.4 m/s. Difference of the propagation velocity for the fourth harmonic will be Δc_4 = 0.6 m/s, i.e., it is at least 8% less.

With account of inverse dependence of dispersion value from the length of surface wave (expression (3.2.1) its effect is stronger for short wave λ = 35 m, nonlinearity and dispersion have the common order – ε = 0.08;

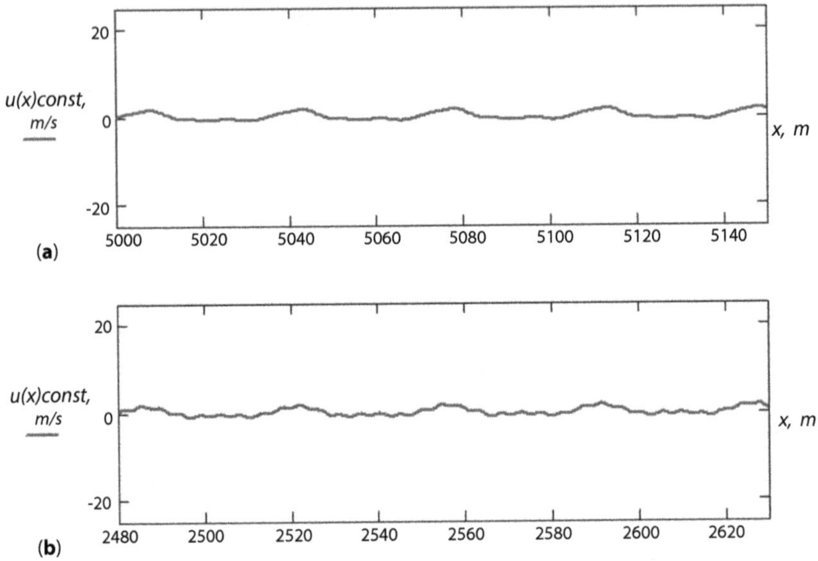

Figure 3.6.2 Surface gravity wave profiles without taking into account (a) and taking into account (b) the dispersion: $f = 0.2$ Hz; $\lambda = 35$ m; $2a/\lambda = 0.024$; $kH = 0.9$; $\varepsilon = 0.08$; $\gamma = 0.02$.

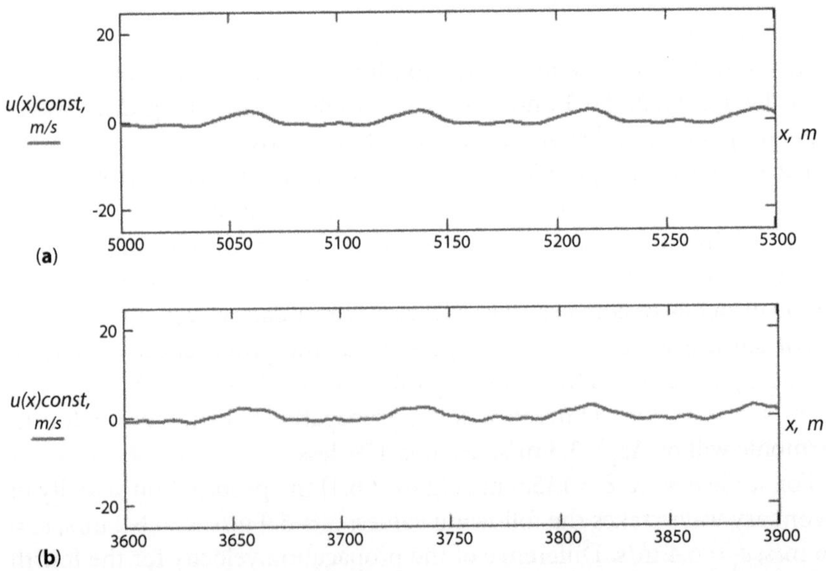

Figure 3.6.3 Surface gravity wave profiles without taking into account (a) and taking into account (b) the dispersion: $f = 0.09$ Hz; $\lambda = 77.8$ m; $2a/\lambda = 0.014$; $kH = 0.4$; $\varepsilon = 0.107$; $\gamma = 0.004$

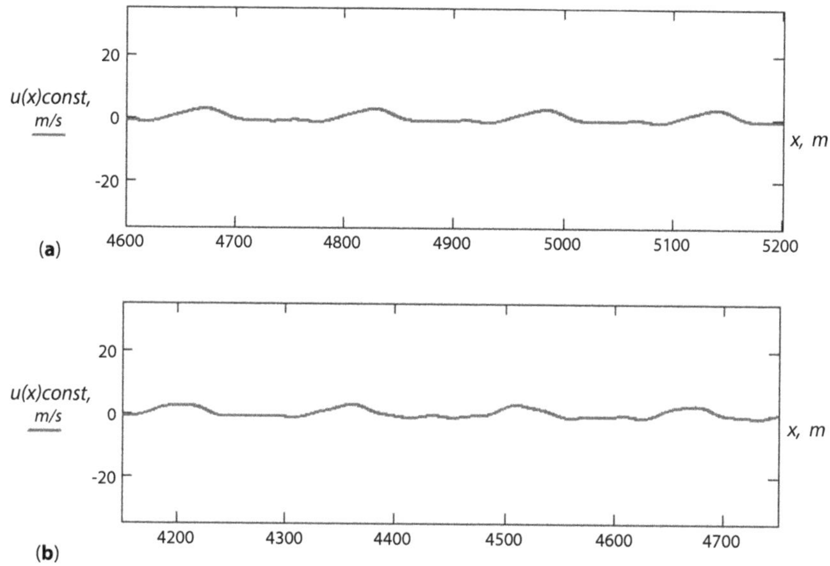

Figure 3.6.4 Surface gravity wave profiles without taking into account (a) and taking into account (b) the dispersion: $f = 0.045$ Hz; $\lambda = 155.6$ m; $2a/\lambda = 0.0088$; $kH = 0.2$; $\varepsilon = 0.136$; $\gamma = 0.001$.

$\gamma = 0{,}02$. For this wave high-frequency fourth harmonic goes over fundamental wave (Figure 3.6.2b).

For surface wave $\lambda = 155{,}6$ m the dispersion occurs in a less degree (Figure 3.6.4), and nonlinearity predominates, it is twice higher than the dispersion – $\varepsilon = 0.136$; $\gamma = 0.001$.

Dispersion results in the loss of synchronism between the secondary waves; and wave profile is no longer periodic. The shallow-water conditions will be better fulfilled and the dispersion will be less notable. Dispersion can also compensate steepening of wave front, and, in an infant state of the propagation, delay its breaking in the bay.

Figure 3.6.5 gives a spatially three-dimensional model of surface gravity wave propagation throughout the bay with and without account of dispersion. Wave parameters are similar to those given in Figure 3.6.3. Wave profile distorts with nonlinearity in Figure 3.6.5a, Figure 3.6.5b gives the wave profile with account of nonlinearity and dispersion. Here we can see wave packet breaking with dispersion.

When approaching the shore, the front of the surface gravity wave will be distorted. For the comparative analysis, the profiles of the surface

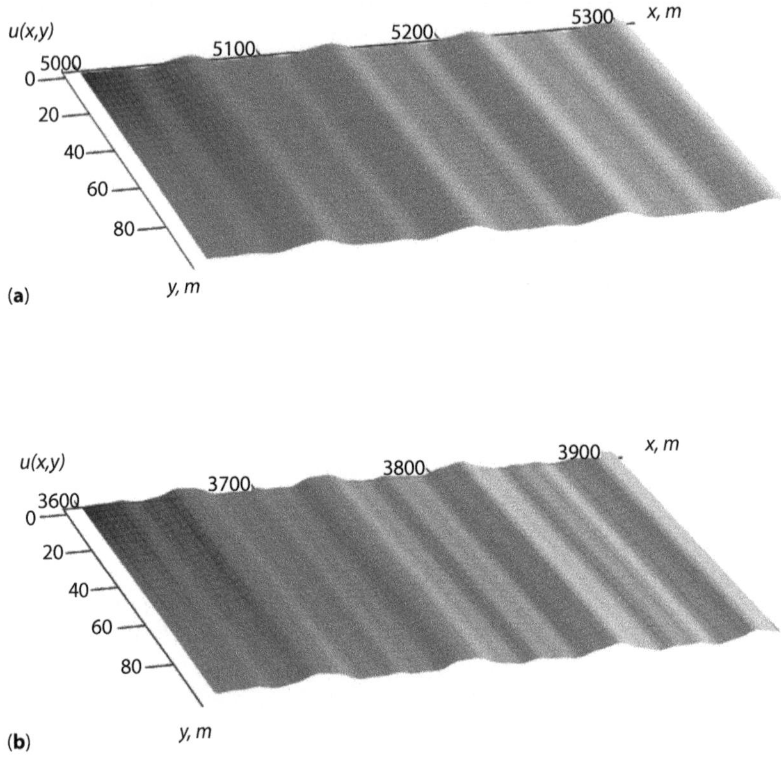

Figure 3.6.5 Spatial model of the propagation of a surface gravity wave with parameters: $f = 0.09$ Hz; $\lambda = 77.8$ m; H = 5 m; $2a/\lambda = 0.014$; $\varepsilon = 0.107$; $\gamma = 0.004$.

gravity wave are disregarded in Figure 3.6.6 without taking into account (Figure 3.6.6a), and taking into account the variance (Figure 3.6.6b). The bottom line is given on the basis of the sine function (Figure 3.5.1).

The instability distance for a surface wave with initial parameters λ = 77.8 m, $2a/\lambda = 0.014$ without dispersion is $x^{(2)}_{char.} \approx 5000$ m, and with dispersion is reduced to $x^{(2)}_{char.} \approx 4300$ m.

Figure 3.6.7 shows a three-dimensional model of the refraction of a nonlinear surface gravity wave on the cape with the dispersion. The length of the bay is from 0 to 4.5 km along the x and y axes. The gravity wave has the following initial parameters: $f = 0.045$ Hz; $\lambda = 155.6$ m; $H = 5$ m; $2a/\lambda = 0.0076$; the approach angle is $\theta = 0^0$ relatively of x axis. With the approach to the cape, the length of the nonlinear gravity surface wave decreases, the front of the wave is distorted. Because of the weak dispersion for a given wave, the wave packet breaks on to the cape.

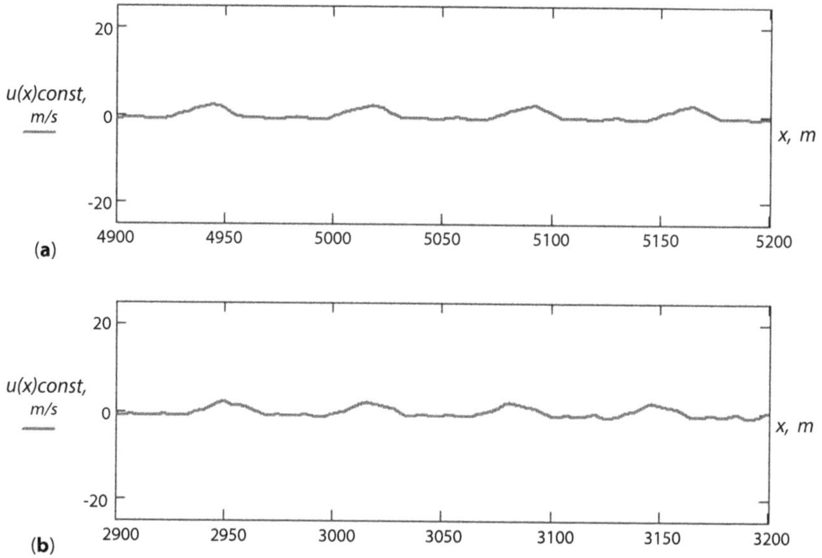

Figure 3.6.6 Refraction profiles of the surface gravity wave without taking into account and taking into account the dispersion, initial parameters: $f = 0.09$ Hz; $\lambda = 77.8$ m; $2a/\lambda = 0.014$; $kH = 0.4$; $\varepsilon = 0.107$; $\gamma = 0.004$.

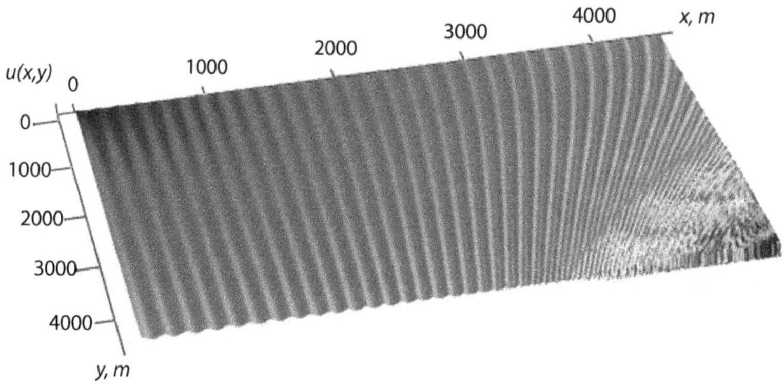

Figure 3.6.7 Three-dimensional model of refraction to the cape of a nonlinear surface gravity wave with allowance for dispersion, initial parameters: $f = 0.045$ Hz; $\lambda = 155.6$ m; $H = 5$ m; $2a/\lambda = 0.0076$; $\theta = 0^0$.

Conclusions

In the third chapter, a study and modeling of nonlinear wave activities on the shallow water surface was carried out. An overview of studies on nonlinear surface gravity waves has been made.

The equations that describe nonlinear waves on the surface of an ideal incompressible fluid in the gravity field were introduced. The equations of the shallow water, also the equations Boussinesq, Korteweg-de Vries, taking into account the effects of variance when extending nonlinear surface waves in shallow waters were considered.

Using the method of successive approximations the equations of shallow water were solved, with regard for quadratic nonlinearity in the second approximation. The solution for the horizontal velocity of particles in the first approximation consists of the main harmonic and the linearly growing second harmonic. The solution in the second approximation contains secondary waves of double frequencies and complex waves. The appearance of a second, third, and fourth harmonic in the spectrum of the source wave over time leads to a distortion of the wave profile.

Based on the energy conservation law, constraints were found for the horizontal velocity of the primary and secondary wave fields. The graphs of the main wave energy change and the higher harmonics from the wave travel were provided. A semigraphical method was used to determine the distance of the instability.

The profiles of nonlinear surface waves for different wave travels were provided. 3D distribution models of the surface gravity wave were provided. The types of asymmetry in surface wave profiles were described.

Wave processes happened at shoaling of nonlinear surface gravity waves. Beach approach of various slopes were modeled. The 3D models of the refraction of a nonlinear surface gravity wave on coastal formations were presented.

In order to adequately consider the distribution of nonlinear surface waves in shallow waters, it is necessary to take into account both nonlinearity and dispersion. Based on the received analytical expressions for the particles horizontal velocity, the nonlinear surface gravity wave profile evolution was described. The nonlinear surface gravity wave profile when the wave is distributed along the creek with the dispersion was introduced. The 3D models of the surface wave in the bay were presented with no taking and taking into account the dispersion. The 3D model of the refraction of a nonlinear surface gravity wave on the cape taking into account the dispersion was provided.

The results of the studies in this chapter have been published in the following main works: [Abbasov, 2010 Fizmatlit], [Abbasov, 2003], [Abbasov, 2004-1], [Abbasov, 2004-2], [Abbasov, 2009], [Abbasov, 2010], [Abbasov, 2011].

4

Numerical Simulation of Nonlinear Surface Gravity Waves in Shallow Water

4.1 Review of Studies on Computational Modeling of Surface Waves

It must be emphasized, the questions of 3D modeling of surface gravity waves, based on analytical solutions of nonlinear dispersion equations of shallow water were examined at work [Abbasov, 2010 Fizmatlit]. Let's consider the results of studies on computational modeling of the wave hydrodynamics of shallow waters:

- based on nonlinear and nonlinear-dispersion equations of shallow water;
- based on the nonlinear-dispersion equations of Boussinesq, Korteweg de Vries;
- based on the Navier-Stokes equations.

Nonlinear and Nonlinear-Dispersion Equations of Shallow Water

Let's describe the works that cover computational modeling of nonlinear and nonlinear-dispersion equations of shallow water. The book [Voltsinger, Piaskowski, 1977] considers the differential methods of integrating the equations of shallow water in some detail. The hydrodynamics of the studied phenomena and the successive conclusion of the main equations and their modifications are described. An overview of the differential schemas for the examined tasks, analysis of their resilience were made. The practical implementation and analysis of computational solutions to some oceanographic problems of shallow water theory were also discussed. Details of all aspects of the computational solution from the creation of the differential grid to the setting of boundary conditions and the assignment of forces were considered.

Computational modeling of wave distribution on the surface of the liquid in a nonlinear-dispersion model (Peregrine model) is considered in the work of Yu. I. Shokin *et al.* [Shokin *et al.*, 1989]. It also considered the task of the run-up of the tsunami wave on the coast. Computational modeling was carried out within the framework of the equations of the shallow water theory of different approximations.

Finite-difference approximation of equations were based on explicit-implicit, finite-difference schemas and were approximated by using the D. Peregrine schema. A free pass condition was provided on the right and left edges of the calculation area [Marchuk An. G. *et al.*, 1982]. Initial disturbance was given on the basis of half-sine.

The work [Marchuk, Moshkalev, 2014] proposes a new method of modeling of long wave run-up to the free-profile coast. Nonlinear equations of shallow water are used to describe the wave distribution to the encroachment line. Behind the encroachment line, a special algorithm is used to define the flow parameters during the wave run-up process.

The results of model calculations without dispersion effects show a monotonic reduction of the maximum heights at any distance from the source, while dispersion models produce a noticeable maximum. Neglecting the nonlinear members from nonlinear-dispersion model reduces the heights of the waves. The solution of the nonlinear-dispersion model is a sequence of soliton-like waves that appeared after the transformation of the initial disturbance. In all cases, a head wave appears pursued by wave train of lower amplitude. In order to evaluate mathematical models, data from field observations and laboratory experiments are used.

The work [Zheleznyak, Pelinovsky, 1985] considers the approximations behind the simplified tsunami run-up models. The interconnections of the various models are discussed, and the dimensionless run-up parameters for processing the results of computational and laboratory experiments are analyzed. The work of [Touboula, Pelinovsky, 2014] is devoted to the study of pressure at the floor when a private wave run-up into a vertical wall. Nonlinear equations take into account dispersion effects. In the work [Didenkulov *et al.*, 2014] the run-up of the bell-shaped pulses of various shapes in the linearly inclined bay of the parabolic section is examined.

In the late 1970s and the early 1980s, there was fairly rapid progress in computational studies of nonlinear surface waves. It should be noted that attempts to computationally solve nonlinear equations raise the question of the adequacy of the discrete presentation. It was noted that sometimes the results of a discrete system study did not correspond to a continuous limit [Bretherton, 1964], [Caponi *et al.*, 1982]. Therefore, the sustainability and conservativeness of a discrete model are important in computational modeling.

The question of transforming monochrome waves over the horizontal floor of the coastal zone is sufficiently studied in the work [Chapalain *et al.*, 1992] and [Eldeberky, Madsen, 1999]. Laboratory experiments and computational modeling examined the influence of the first four harmonics on the profile of the wave when it was distributed in shallow waters. The system of evolutionary equations was solved numerically by the Runge-Kutta method. The model was verified and tested according to laboratory and in-kind experiments. The special aspects of the process of distorting the wave profile in the distribution process.

Numerical simulation of nonstationary periodic surface waves is carried out in the article by V. R. Kogan and V. V. Kuznetsov [Kogan, 1995]. The problem of the evolution of surface waves reduces to solving a system of ordinary differential equations. Dynamics of gravity waves at various amplitudes and lengths of waves is numerically simulated. The profiles of gravity waves at the initial time and at the time of breaking are given.

Simulation of breaking waves in a fluid of finite depth is described in the work of V. A. Kalmykov [Kalmykov, 1996] uses a three-wave Fourier-wave-amplitude equation for this. The waves acquire a pronounced asymmetry in the direction of motion. The presence of asymmetry relatively to the vertical axis was experimentally observed in the work of O. Phillips [Phillips, 1966], the formation of an asymmetric wave in the direction of motion is given in Longuet-Higgins, 1993. The problems of wave breaking were described in detail in a review by D. Peregrine [Peregrine, 1993].

The article by A. A. Litvinenko and G. A. Khabakhpashev [Litvinenko, 1999] is devoted to numerical modeling of nonlinear long waves in basins with a shallow bottom. A nonlinear dispersion model of shallow water is considered taking into account the topography and the viscosity of the liquid. A comparison is made of the calculations of the transformation of a plane disturbance of the free water surface with experimental data. The problem of the effect of a conical-cylindrical island, an underwater ridge and friction on an inclined bottom on the evolution of a flat solitary wave is solved numerically.

The problems of modeling spatial nonlinear waves are considered in the work of D. G. Arkhipov and G. A. Khabakhpashev [Arkhipov, 2006]. The dynamics of finite amplitude waves traveling in different directions in the presence of dispersion is described.

A finite-difference algorithm for modeling surface waves within of a nonlinear dispersion model is considered in [Barakhnin, Khakimzyanov, 1996]. A distinctive feature of the algorithm is the selection in the initial equations of the elliptic and hyperbolic parts. To solve the resulting elliptic equation, a finite-difference scheme with a self-adjoint and positive definite operator is constructed, bounds of the spectrum of this operator are estimated.

In [Khakimzyanov, Shokina, 1996], a finite-difference method is described for calculating steady flows of an ideal incompressible fluid with surface waves in channels with a complex outline of the shores. The questions of constructing finite-difference approximations on curvilinear non-orthogonal meshes for describing the geometry of a reservoir are discussed. The results of calculations of the problem of the steady flow of a liquid with surface waves in a complex reservoir are presented.

In [Vinje, Brevik, 1981], a numerical simulation of the evolution of the profile of a breaking surface wave was carried out. The wave profiles are constructed on the basis of the solution of the system of equations (from the Laplace equation) by numerical methods.

The paper [Hodgins *et al.*, 1985] is devoted to numerical modeling of the transformation of a surface wave in shallow water. A review of the work on numerical modeling and practical applications in Canadian water reservoirs is given. The processes of refraction of surface waves of finite amplitude in shallow water are described. However, the theory of finite-amplitude waves differs significantly from the linear theory in predicting the altitudes and the refractive index of the surface wave. Nonlinear waves are refracted less than linear waves.

The paper [Elgar *et al.*, 1998] is devoted to numerical simulation and experimental observations of the effects of nonlinear interaction, reflection, and damping effects on the propagation of surface gravity waves in the coastal zone. Nonlinear interactions lead to doubling the number of crests of the wave, but for waves of a lower height the crest did not bifurcate. The described effects are considered within of the Boussinesq model.

In the work [Afanasyev, Stukolov, 2013] a numerical model of the work of an experimental wave-generator of single waves is presented. The results of a comparison of the experimental and numerical modeling of the transformation of single waves on a shallow slope are given.

In the article by N. A. Kudryashov *et al.* [Kudryashov, 2003], mathematical modeling of gravity waves is carried out in the shallow-water approximation. The motion of a solitary wave along the surface of water, the wave output to shore, and the passage of a solitary wave over an underwater rock are investigated. When modeling the motion of a soliton wave, the approximation of shallow water begins to be violated at a ratio of the water depth to the wavelength equal to 0.3. The approximation of shallow water can be used to estimate the distance from the shore on which the wave is breaking.

The work [Fedotova, 2002] is devoted to the finite-difference method of a through account, using a rectangular grid. The proposed method is suitable for solving a wide class of problems of wave hydrodynamics in pools with arbitrary bottom relief and complex outlines of the coastline.

The paper [Fedotova, 2004] considers a system of equations describing the propagation of long surface waves on water with allowance for dispersion effects. Such models are used to solve the problems of long-wave hydrodynamics, when the use of complete three-dimensional models leads to unreasonably large expenditures of computing resources.

In [Fedotova, Khakimzyanov, 2010], nonlinear dispersion equations of shallow water are obtained on the surface of a sphere. Equations can be used to simulate the propagation of tsunami waves over long distances, taking dispersion into account.

In the work [Beyzel *et al.*, 2010] the problem of numerical modeling of surface waves generated by the motion of a flooded landslide along a spatially inhomogeneous slope is considered. The law of motion of a landslide under the action of gravity, friction, ejection and hydrodynamic resistance of the surrounding fluid is presented. The results of computational experiments for the model water area are discussed.

Let us also consider the works devoted to modeling the hydrodynamics of the Azov Sea on the basis of shallow water equations, since we use the hydrophysical conditions of the Azov Sea as a shallow water model.

The mathematical modeling of hydrodynamics of the Azov Sea is devoted to the work of L. A. Krukier, A. L. Chikin, G. V. Muratova. The work of L. A. Krukier [Krukier, 1991], [Krukier *et al.*, 2008] is devoted to an assessment of the change in the hydrodynamics of the Sea of Azov in the case of projects for the reconstruction of its ecosystem. For this, a mathematical model based on a numerical solution of the complete system of nonlinear shallow water equations is used. In the works [Chikin, 2001], [Muratova, 2004], [Krukier *et al.*, 2009] describes the three-dimensional problem of determining the hydrodynamic flow parameters in the Azov Sea and the Taganrog Bay, and flow patterns for various wind situations are obtained. The basis of these hydrodynamic models is the system of equations for the theory of shallow water with nonlinear terms, obtained in the work of N. E. Voltsinger and R. V. Piaskovsky [Voltsinger, 1977].

The construction of high-precision three-dimensional models of hydrodynamics of the Azov Sea is devoted to the work of A. I. Sukhinov. The works [Sukhinov, 1987, 2003, 2007] are devoted to the construction of high-precision 2D and 3D models of hydrodynamics, heat transport, salts and suspensions in shallow water bodies. Physico-mathematical models of turbulent exchange in marine systems have been created, methods for solving grid equations of mathematical physics for mass parallel computing systems have been constructed. In the articles [Sukhinov, 2003], [Debol'skaya, Sukhinov, 2005] the results of complex oceanological studies of the Sea of Azov are presented, the mechanisms of the appearance of giant natural traps accumulating pollution and suspended matter are explained.

It should also be noted that some papers devoted to the numerical calculation of discontinuous waves on the basis of the shallow water equation.

The paper [Delis, Skeels, 1998] is devoted to the numerical solution of shallow water equations using implicit finite difference schemes for modeling discontinuous waves. A rather detailed review of the work on the numerical solution of the shallow water equations is presented in the monograph [Kulikovsky *et al.*, 2001]. Basically, these solutions are designed to simulate one-dimensional and two-dimensional fluid flows with medium and large Froude numbers.

In [Viottia *et al.*, 2014], a study is made of the run-up of nonlinear high-steepness waves onto a vertical wall, taking into account nonlinear-dispersion phenomena.

A numerical algorithm for simulating the propagation of discontinuous waves along the dry bed based on shallow water equations is proposed [Borisova, Ostapenko, 2006]. The results of numerical simulation of the propagation of a discontinuous wave arising from the destruction of a dam in a channel with an uneven bottom are presented.

In [Kushnir *et al.*, 2013] the problem of changing the structure of surface waves and the formation of shock loads during their destruction in the immediate vicinity of shore structures is considered.

Nonlinear-Dispersion Equations of Boussinesq and Korteweg and de Vries

We describe some studies devoted to the numerical simulation of surface waves on the basis of the nonlinear dispersion equations of Boussinesq and Korteweg and de Vries.

The paper [Elgar *et al.*, 1998] is devoted to numerical simulation and experimental observations of the effects of nonlinear interaction, reflection, and damping effects on the propagation of surface gravity waves in the coastal zone. Nonlinear interactions lead to doubling the number of crests of the wave, but for waves of a lower height the crest did not bifurcate. The described effects are considered within of the Boussinesq model.

[Lee *et al.*, 1998] presents theoretical and numerical studies of the two-soliton solution of the Korteweg-de Vries equation of the second kind. The results show that there exists a solution in the form of a "quasisoliton" for the given equation. In this solution, solitary waves retain their similarity to the pure solitons of the Korteweg-de Vries equation.

The paper [Elgar *et al.*, 1998] is devoted to the study of the spectrum of a nonlinear surface wave in shallow water within of the Boussinesq stochastic model. Stochasticity is based on the phase relationship between the frequency components of the wave. With the help of numerical simulation, the general features of the stochastic model are illustrated, depending on the nonlinearity, on the initial form of the spectrum.

A comparison of studies based on Boussinesq models with experimental observations of collapsing surface gravity waves in shallow water was carried out in [Chen *et al.*, 1997]. The change in the spectral composition of the wave in the surf zone is modeled by a more accurate decomposition. Despite the special extension of the Boussinesq model to the region of breaking waves, the proposed models describe the observed change in the spectrum, as well as the asymmetry of the wave profile.

Within of the Boussinesq model, the nonlinear dispersion of surface waves in shallow water is considered in the article [Herbers *et al.*, 2002]. The linear model is obtained as a first-order solution, in the second approximation the model contains a weak dispersion of frequency and amplitude. Experimental data on surface waves for an inclined sandy beach are presented. The nonlinear Boussinesq theory describes well the irrotational motion of a wave within the surf zone.

The work [Shermeneva, 2006] is devoted to mathematical modeling of the propagation of nonlinear waves on the water surface. A method for the analytical solution of the Boussinesq equations of high order is proposed. With the help of the solutions obtained, the surface waves are run-up on a sloping shore. Expressions describing the transformation of surface waves over an obstacle with elliptical level lines are derived. It is shown that a wave pattern on the surface can be used to judge the shape, location and orientation of the obstacle.

The search for discontinuous solutions of nonlinear-dispersion equations is carried out in [Shimozono, Sato, 2009]. As initial equations, Boussinesq type equations with different sets of parameters are used. The search for the solution is based on a numerical calculation of the profile of the breaking wave. It is noted that solutions exist for waves under the condition $h/L<0.04$. The paper presents distortions of the surface wave profile with different values of the nonlinearity parameter.

In [Gusev *et al.*, 2014] investigated the dispersion effect for the picture surface waves encountered in gathering underwater landslide in a confined body of water. To describe surface waves, both the full nonlinear dispersion model of shallow water and the new approximate Boussinesq models are used.

The Navier-Stokes Equations

The analytical integration of the Navier-Stokes equations in the literature is very rare [Lamb, 1947], [Loitsiansky 1978]. Therefore, progress in this area is possible mainly through the use of numerical methods. At present, several solutions of varieties of difference schemes are used for the numerical solution of the Navier-Stokes equations.

Most of them were developed with reference to the system of equations in the variables vortex-current function [Roach, 1980]. A common drawback of this approach is the need to use the boundary condition for a vortex on a solid surface. Also, the use of the variables vortex-current function excludes the possibility of generalizing this method to spatial and turbulent flows. This causes an increased interest of the researchers in solving the Navier-Stokes equations in physical velocity-pressure variables. Such an approach allows solving both the two-dimensional and three-dimensional problems by a single algorithm. However, this path is associated with the difficulties in calculating the pressure field consistent with the velocity field.

The Navier-Stokes equations have a number of specific features that significantly affect their numerical solution. One of the essential features is the nonlinearity and the parabolic-elliptic nature of these equations. Therefore, for an adequate simulation of the elliptic nature of the Navier-Stokes equations, it is necessary to use the elliptic Poisson equation for pressure. The presence of a small parameter with the highest derivative and the nonlinearity of the Navier-Stokes equations in combination with their ellipticity (for large Reynolds numbers) create the conditions for the formation of complex spatio-temporal vortex structures in the liquid. This leads to nonstationarity of the liquid flow, loss of its stability, and transition to turbulent flow regime.

One of the first numerical methods for solving the Navier-Stokes equations for an incompressible fluid in natural physical velocity-pressure variables was proposed by Chorin (1967) and became known as the method of "artificial compression". It is based on the fact that a term with an artificial compressibility is added to the equation of noncontinuity, which vanishes when the solution is established in time. Chorin also proposed a splitting method for integrating the system of Navier-Stokes equations, the essence of which consists in dividing the described physical process into two: convection-diffusion and the contribution of pressure [Chorin, 1968].

Then other methods were proposed. Among them are the method of "variable directions", developed by Douglas [Douglas, Gunn, 1964], and Peaceman, Rachford [Peaceman, Rachford, 1955]. The method is based on using the reduction of a multidimensional problem to a sequence of one-dimensional problems with tridiagonal matrices.

Of the most early and well-known methods of solution is the method of "markers and cells" (MAS), proposed in the works of Harlow [Harlow, Welch, 1965], [Harlow, 1967]. Later this method was constantly modified. It is characterized by the fact that the original equations are stored in the velocity-pressure variables, and a mesh is used to construct the finite-difference scheme.

The other part of the methods is related to the splitting methods considered in the book of G. I. Marchuk [Marchuk, 1988]. Among them, we can mention the method of splitting by the physical processes of Belotserkovsky [Belotserkovsky, 1994]. Further on the basis of the ideas of the MAC method, O. M. Belotserkovsky and Yu. M. Davydov proposed the "large particle method" [Belotserkovsky, Davydov, 1971].

A single-step and two-step scheme of Lax-Wendroff and McCormack [Richthmaier, Morton, 1972], [Fletcher, 1991], which consists of the use of the "predictor-corrector" scheme, was widely used in the problems of hydrodynamics. In time, a number of methods were proposed in a number of works, which were called the SIMPLE methods. Among them, we can mention the work [Patancar, Spolding, 1972]. A detailed review of the solution methods is presented in the work [Bystrov *et al.*, 2005].

At present, in connection with the great practical importance, the search for effective difference schemes and algorithms for solving the equations of motion of a viscous incompressible fluid extends intensively. The evaluation of their quality, respectively, is determined by comparing the results of calculations with other known calculated and experimental data.

We describe the studies devoted to the numerical modeling of surface waves on the basis of the nonlinear Navier-Stokes equations. The Navier-Stokes equation makes it possible to describe not only nonlinear effects, but also makes it possible to take into account turbulent processes in a viscous incompressible fluid. Therefore, the Navier-Stokes equation, along with hydrodynamic problems, is often used to describe the onset of surface waves on the coastal slopes of shallow water areas.

Modeling of the problems of wave hydrodynamics on the basis of the MAC method was carried out in [Amsden, Harlow, 1970]. [Hirt, Nichols, 1981] presents a finite-volume method (VOF, volume of fluid) with density markers (in replacement of the particle-marker method).

The development of methods for the numerical integration of complete hydrodynamic equations in the presence of a free surface is devoted to the work [Belotserkovsky *et al.*, 1987]. With the help of the proposed effective numerical method, both steady-state and non-stationary wave motions of a liquid are investigated. The proposed version of the method of splitting by physical factors for the calculation of incompressible fluid flows makes it possible to significantly reduce computer time by developing an effective algorithm for solving a boundary value problem for an elliptic-type equation.

The paper [Kawasaki, 1999] is devoted to the investigation of a two-dimensional numerical model of the effect of a flooded breakwater on wave propagation. The two-dimensional model is constructed on the basis of the Navier-Stokes equations. The problems of wave modeling are considered both before breaking and after breaking. To test the model, the results of laboratory experiments are presented. The transformation of the wave profile is analyzed, and the relation between the steepness of the wave and its spectral composition. It is shown that the proposed numerical model can well reproduce the deformation processes of the wave profile before and after breaking.

[Adalsteinsson, Sethian, 2002] describes a two-dimensional numerical model of the wave breaking in waters with inclined coastal slopes. Modeling uses a nonlinear potential model of an incompressible, irrotational, inviscid flow taking into account the topography of the bottom. The speed calculation algorithm is based on the Boundary Element Method (BEM) method.

The work [Liu, Losada, 2002] is devoted to a two-dimensional numerical simulation of wave propagation from the deep-water zone to the surf zone, taking into account its breaking. We are looking for a combined method that takes into account both deep-sea nonlinear dispersion models and the method based on Reynolds-Averaged Navier-Stokes (RANS) for simulating breaking waves. Two-dimensional models of wave breaking on a coastal breakwater are presented.

In the work [Colagrossi, Landrini, 2003] numerical simulation of a two-dimensional wave based on the method of Smoothed Particle Hydrodynamics (SPH) is carried out. Stages of stiffening, breaking and further spread of wave in basins with variable and constant depths are presented and studied. The formation of vortices on the front slope of a wave during the breaking and its effect on shore erosion are described.

The paper [Zhao *et al.*, 2004] is devoted to the economical two-dimensional turbulent model of breaking waves. The free surface of the liquid is traced by the MAC method based on the RANS method. The paper also discusses the influence of the scale of the sublattice on the mechanism of turbulent transport. The profiles of breaking waves are compared on the basis of various methods for propagation along a sloped bottom.

In the work [Bakhtyar *et al.*, 2009], a two-dimensional simulation of the wave run-up process on an inclined shore is performed taking into account the two-phase state of the air-water flow. For modeling, the Reynolds averaging method of the Navier-Stokes equations (RANS) is used.

The work [Kshevetskii, Vereshchagina, 2014] is devoted to a two-dimensional modeling of the tidal wave of a tsunami taking into account a two-layer water-air model. The effect of the bottom slope of the coastal zone on the breaking of the wave is estimated.

[Kawasaki *et al.*, 2010] presents a two-dimensional wave impingement model based on a multiphase model of solid-gas-liquid interaction. By the method of splitting, the initial equations are divided into advective and non-advective parts. At the advective stage, the CIP method calculates the hyperbolic equations for velocity and pressure. At the non-advective stage, the equations are solved by the extended SMAC method. In this paper, the process of deformation of the profile of a wave is described in the course of a run to an inclined shore. Processes are simulated, both before and after the breaking of the wave.

Three-Dimensional Models of Wave Hydrodynamics

Let us consider the works devoted to the three-dimensional modeling of the run-up of surface wave on shore formations.

The articles [Lubin *et al.*, 2006], [Watanabe, Saeki, 1999] are devoted to three-dimensional modeling of the breaking of the surface wave. The paper [Lubin *et al.*, 2006] presents the results of three-dimensional modeling of run-up, and breaking of the wave during propagation along an sloping shore. For modeling, the Navier-Stokes equations are used, taking into account the two-phase air-water environment. The breaking of the wave is considered taking into account the processes of sharpening, stiffening, overturning, bursting upwards and capturing the air mass. A comparison of the simulation results with the experimental measurements presented in [Kimmoun, Branger, 2007] is made.

In [Ting, Kirby, 1996], experimental studies of the deformation of the profile of breaking surface waves in the surf zone are also presented. The measurements for the horizontal and vertical components of the velocity vector are presented. Staged photographs of deformation of the wave profile during propagation along a sloped bottom before and after breaking are given.

In [Fedotova, 2002], a numerical method for solving the long-wave coastline problem is investigated. A finite-difference count-through method is described that uses a rectangular grid in a fixed region. The proposed method can be suitable for solving a wide class of problems of wave hydrodynamics in pools with arbitrary bottom relief and complex outlines of the coastline. The method allows solving two-dimensional problems with complex geometry. A three-dimensional fragment of a splash of a tsunami on the real coast is represented.

The paper [Lubin *et al.*, 2004] is devoted to three-dimensional modeling of the wave breaking process on the basis of the Navier-Stokes equations taking into account the turbulences of the air and water environments. Particular attention is paid to the step of capturing the air mass and the formation of vortex flows with foam and bubbles.

In [Borisova, 2007], within the theory of shallow water, a method is proposed for simulating the onset of discontinuous waves on an inclined shore and their propagation along a dry bed. A three-dimensional model of the problem of partial destruction of a dam with a local ellipsoidal obstacle is presented.

In [Delis, Kazolea, 2008], the propagation of storm waves and tsunami is simulated within a numerical solution of the nonlinear shallow water equation. A step-by-step three-dimensional model of the tsunami downwash into coastal environment is presented.

The paper [Wu Y.-T. *et al.*, 2014] is devoted to the three-dimensional modeling of the effect of solitary waves on coastal breakwaters. The paper [Haa *et al.*, 2014] deals with 3D modeling of the rolling of a solitary wave onto a circular island based on the Navier-Stokes equations. When creating a grid, bottom structures of complex shape are taken into account.

As a result of the review, it can be noted that there are many works devoted to numerical modeling of wave processes both on the basis of shallow water equations and the Navier-Stokes equations.

When modeling on the basis of the shallow water equation, iterative methods for solving discrete equations are used in most sources. In our case, we will use the exact methods for solving discrete equations. Numerical simulation within the theory of shallow water also allows us to verify the correctness of the results of modeling on the basis of the developed graph-analytic method for describing the transformation of surface gravity waves.

Within our research, we will numerically simulate two-dimensional and three-dimensional problems in the approach of surface waves to coastal slopes. For modeling, we will mainly be attached to the hydrophysical conditions of the Azov Sea. These issues remain quite relevant for predicting the power effects of wave processes on shore formations.

Figure 4.1.1 and 4.1.2 present the author's photo illustrations characterizing the stages of breaking of surface gravity waves in the surf zone.

Figure 4.1.1 Stages of breaking of surface waves in the conditions of the bay (*Dyurso village, the Black Sea*).

Figure 4.1.2 Stages of breaking of surface waves in the conditions of the bay (*Dyurso village, the Black Sea*).

4.2 Statement of the Problem

Construction of a Discrete Model

Let us turn to the numerical solution of the shallow water equations. The system of shallow water equations contains a continuity equation and dynamic equation on the conservation of momentum:

$$\begin{cases} \dfrac{\partial u}{\partial t} + u\dfrac{\partial u}{\partial x} = -g\dfrac{\partial \zeta}{\partial x} \\ \dfrac{\partial \zeta}{\partial t} = -\dfrac{\partial (hu)}{\partial x} \end{cases} \tag{4.2.1}$$

where $h = H + \zeta$ – distance from bottom to surface of liquid, u is medium particle velocity, ζ – surface swell function, H – fluid depth. Shallow water equations ignore the dispersion phenomenon as it is of minor importance in shallow water.

Boundary conditions. The *kinematic* boundary condition is taken as a boundary condition on free liquid surface, i.e., surface swelling velocity is the same as medium particle vertical velocity. To follow the *dynamic* condition, the free liquid surface pressure is considered the same as the atmospheric pressure. As a bottom condition, vertical intensity of liquid particle velocity is assumed to be equal to zero (sticking condition):

$$\frac{\partial\zeta}{\partial t}+u\frac{\partial\zeta}{\partial x}=w\Big|_{z=\zeta(x,t)},\ p\Big|_{z=\zeta(x,t)}=g\rho_0\zeta,u\Big|_{z=-H}=0. \qquad (4.2.2)$$

Splitting method. The discrete analogue of the difference derivative in the time will have the form:

$$\frac{\partial u}{\partial t}=\frac{\hat{u}-u}{\tau}. \qquad (4.2.3)$$

For the first equation of the system (4.2.1) (the equation of the law of conservation of the volume of a liquid), we will perform an approximation in the time coordinate:

$$\frac{\hat{u}-u}{\tau}+u\frac{\partial u}{\partial x}=-g\frac{\partial\zeta}{\partial x}. \qquad (4.2.4)$$

Let us apply the method of *splitting* by physical processes to this equation

$$\begin{cases}\dfrac{\tilde{u}-u}{\tau}+u\dfrac{\partial u}{\partial x}=\mu\dfrac{\partial^2 u}{\partial x^2}\\ \dfrac{\hat{u}-\tilde{u}}{\tau}=-g\dfrac{\partial\zeta}{\partial x}\end{cases} \qquad (4.2.5)$$

where u– a velocity component on the current time layer;
$\tilde{u}$ – a velocity component on the auxiliary time layer;
$\hat{u}$ – a velocity component on the next time layer;
τ – a time step, μ – a lattice velocity parameter.

Let us multiply the second system equation (4.2.5) by h and take the derivative in x- time coordinate of both members:

$$\frac{1}{\tau}\left(\frac{\partial(h\hat{u})}{\partial x}-\frac{\partial(h\tilde{u})}{\partial x}\right)=-\frac{\partial}{\partial x}\left(hg\frac{\partial\zeta}{\partial x}\right). \tag{4.2.6}$$

Multiply both sides by the time step t

$$\frac{\partial(h\hat{u})}{\partial x}-\frac{\partial(h\tilde{u})}{\partial x}=-\tau\frac{\partial}{\partial x}\left(hg\frac{\partial\zeta}{\partial x}\right). \tag{4.2.7}$$

Let us write the second equation of the system (4.2.1) (the law of conservation of momentum) in the form:

$$\frac{\partial\zeta}{\partial t}=-\frac{\partial(h\hat{u})}{\partial x}. \tag{4.2.8}$$

Then equation (4.2.7) takes the form:

$$\frac{\partial\zeta}{\partial t}+\frac{\partial(h\tilde{u})}{\partial x}=\tau\frac{\partial}{\partial x}\left(hg\frac{\partial\zeta}{\partial x}\right). \tag{4.2.9}$$

taking into account the expression for $h=H+\zeta$:

$$\frac{\partial\zeta}{\partial t}+\frac{\partial(\tilde{u}\zeta)}{\partial x}=\tau g\frac{\partial}{\partial x}\left((\zeta+H)\frac{\partial\zeta}{\partial x}\right)-\frac{\partial(\tilde{u}H)}{\partial x}. \tag{4.2.10}$$

The expression is a constraint equation of level swelling function ζ and velocity component $\tilde{u}$ on the auxiliary time layer.

Thus the system of equations (4.2.1) changes to:

$$\begin{cases}\dfrac{\tilde{u}-u}{\tau}+u\dfrac{\partial u}{\partial x}=\mu\dfrac{\partial^2 u}{\partial x^2}\\[2ex] \dfrac{\partial\zeta}{\partial t}+\dfrac{\partial(\tilde{u}\zeta)}{\partial x}=\tau g\dfrac{\partial}{\partial x}\left((\zeta+H)\dfrac{\partial\zeta}{\partial x}\right)-\dfrac{\partial(\tilde{u}H)}{\partial x}\\[2ex] \dfrac{\hat{u}-\tilde{u}}{\tau}=-g\dfrac{\partial\zeta}{\partial x}\end{cases} \tag{4.2.11}$$

The operational algorithm of the system of equations (4.2.11) is the following:

- from the first equation one finds the components on the auxiliary time layer $-\tilde{u}$, through velocity components on the current time layer $-u$;
- then, from the second equation one finds level swelling function – ζ;
- finally, from the third equation one finds velocity components on the next time layer $\hat{u}$.

The task is divided into three stages; at the first stage:

$$\frac{\tilde{u}-u}{\tau}+u\frac{\partial u}{\partial x}=\mu\frac{\partial^2 u}{\partial x^2}. \tag{4.2.12}$$

$u(x,0)=0$ – at the time zero the initial condition is assumed to be, at any other time the change in the surface form is given according to the harmonic law $u(0,t)=a\sin(\omega t)$, where a, w – surface wave amplitude and circular surface wave frequency.

At the second stage: finding the displacement function relatively to the equilibrium state through the velocity

$$\frac{\partial\zeta}{\partial t}+\frac{\partial(\tilde{u}\zeta)}{\partial x}=\tau g\frac{\partial}{\partial x}\left((\zeta+H)\frac{\partial\zeta}{\partial x}\right)-\frac{\partial(\tilde{u}H)}{\partial x}, \tag{4.2.13}$$

$\zeta(x,0)=0$ – the initial displacement is zero on the whole straight line.

The third stage: correction to the velocity. The velocity components through the components on the auxiliary layer are explicitly expressed taking into account the displacement function:

$$\frac{\hat{u}-\tilde{u}}{\tau}=-g\frac{\partial\zeta}{\partial x}. \tag{4.2.14}$$

Extra viscosity factor μ ($0<\mu<1\text{m}^2/\text{s}$) has been introduced into the system of shallow water equations (4.2.11) when splitting into physical processes. Shallow water equations are hyperbolic systems of equations. Nonlinear hyperbolic equations, compared to linear equations, possess a number of vital differences. These differences should not be overlooked in their numerical integration. Even if the initial conditions are as smooth as desired, the solution of nonlinear equations can contain discontinuities.

In order to avoid this difficulty in practical solving of the problems of nonlinear mechanics, a small additional perturbation in the form of artificial viscosity is introduced into the differential system (similar to Von Neumann-Richtmyer Method) [Richtmyer, 1967], [Woltsinger *et al.*, 1977]. This viscosity eliminates discontinuities and leads to adequate results.

Discrete Model Construction

A difference scheme is a finite system of algebraic equations that corresponds to some differential problem containing a differential equation and additional conditions. Thus, difference schemes are used to reduce the differential problem to a finite system of equations, the numerical solution of which is possible in principle on computers [Bakhvalov *et al.*, 2006].

To construct a solution of the difference scheme, we use a uniform grid [Samarsky, 1989]:

$$\varpi_h = \left\{ t^n = n\tau,\, x_i = ih,\ N_t\tau = T,\ i = 1\ldots N_x - 1,\ N_x h = l \right\}. \quad (4.2.15)$$

where t– is a step in time, h– a step in space, N_t– an upper bound on time, N_x– a border in space.

We use the integro-interpolation method (balance method) to construct the difference scheme of the problem (4.2.12) - (4.2.14). The implicit scheme has been chosen for its greater stability factor.

To do this, we integrate twice the time equations from t_j before t_{j+1} and over the space from $x_{i-1/2}$ to $x_{i+1/2}$ on the grid (4.2.15).

For the equation (4.2.11) we obtain:

$$\int_{t_j}^{t_{j+1}} \int_{x_{i-\frac{1}{2}}}^{x_{i+\frac{1}{2}}} \frac{\partial u}{\partial t} dxdt + \int_{t_j}^{t_{j+1}} \int_{x_{i-\frac{1}{2}}}^{x_{i+\frac{1}{2}}} u \frac{\partial u}{\partial x} dxdt = \int_{t_j}^{t_{j+1}} \int_{x_{i-\frac{1}{2}}}^{x_{i+\frac{1}{2}}} \mu \frac{\partial^2 u}{\partial x^2} dxdt \quad (4.2.16)$$

After integrating each term, equation (4.2.16) is transformed to the form

$$\frac{u_i^{j+\frac{1}{2}} - u_i^j}{\tau} + \frac{1}{2}\left(\frac{u_{i+1}^j + u_i^j}{2} \cdot \frac{u_{i+1}^{j+\frac{1}{2}} - u_i^{j+\frac{1}{2}}}{h_x} + \frac{u_{i-1}^j + u_i^j}{2} \cdot \frac{u_i^{j+\frac{1}{2}} - u_{i-1}^{j+\frac{1}{2}}}{h_x} \right) =$$

$$= \mu\left(\frac{u_{i+1}^{j+\frac{1}{2}} - u_i^{j+\frac{1}{2}}}{h_x^2} - \frac{u_i^{j+\frac{1}{2}} - u_{i-1}^{j+\frac{1}{2}}}{h_x^2}\right). \qquad (4.2.17)$$

Approximations of the initial condition and the boundary condition of the first kind will have the form:

$$u(x,0) = 0 \rightarrow u_l^0 = 0; u(0,t) = a\sin(\omega t) \rightarrow u_0^j = a\sin(\omega t^j).$$

We obtained a discrete analogue for the equation (4.2.12) on the grid (4.2.15). Similarly integrate twice equation (4.2.13) over time from t_j to t_{j+1} and from the space $x_{i-1/2}$ to $x_{i+1/2}$ the grid (4.2.15):

$$\int_{t_j}^{t_{j+1}}\int_{x_{i-\frac{1}{2}}}^{x_{i+\frac{1}{2}}} \frac{\partial \zeta}{\partial t} dxdt + \int_{t_j}^{t_{j+1}}\int_{x_{i-\frac{1}{2}}}^{x_{i+\frac{1}{2}}} \frac{\partial(\tilde{u}\zeta)}{\partial x} dxdt =$$
$$\int_{t_j}^{t_{j+1}}\int_{x_{i-\frac{1}{2}}}^{x_{i+\frac{1}{2}}} \tau g \frac{\partial}{\partial x}\left((\zeta + H)\frac{\partial \zeta}{\partial x}\right) dxdt - \int_{t_j}^{t_{j+1}}\int_{x_{i-\frac{1}{2}}}^{x_{i+\frac{1}{2}}} \frac{\partial(\tilde{u}H)}{\partial x} dxdt \qquad (4.2.18)$$

After integration, the expression (4.2.18) takes the form:

$$\frac{\left(\zeta_i^{j+1} - \zeta_i^j\right)}{\tau} + \left(\frac{\tilde{u}_{i+1}\zeta_{i+1}^{j+1} - \tilde{u}_{i-1}\zeta_{i-1}^{j+1}}{2h_x}\right) =$$
$$\tau g\left(\left(\frac{\zeta_{i+1}^j + H_{i+1} + \zeta_i^j + H_i}{2}\right)\cdot\frac{\zeta_{i+1}^{j+1} + \zeta_i^{j+1}}{h_x^2} - \left(\frac{\zeta_i^j + H_i + \zeta_{i-1}^j + H_{i-1}}{2}\right)\cdot\frac{\zeta_i^{j+1} + \zeta_{i-1}^{j+1}}{h_x^2}\right) - \qquad (4.2.19)$$
$$-\left(\frac{\tilde{u}_{i+1}H_{i+1} - \tilde{u}_{i-1}H_{i-1}}{2h_x}\right).$$

Expression (4.2.19) is a discrete analog of equation (4.2.13) on the grid (4.2.15). Approximation of the initial condition $\zeta(x,0) = 0 \rightarrow \zeta_l^0 = 0$.

We integrate equation (4.2.14) in time from t_j up to t_{j+1} and over the space from $x_{i-1/2}$ to $x_{i+1/2}$ on the grid (4.2.15):

$$\int_{t_j}^{t_{j+1}}\int_{x_{i-\frac{1}{2}}}^{x_{i+\frac{1}{2}}}\frac{\partial u}{\partial t}dxdt=-g\int_{t_j}^{t_{j+1}}\int_{x_{i-\frac{1}{2}}}^{x_{i+\frac{1}{2}}}\frac{\partial \zeta}{\partial x}dxdt. \tag{4.2.20}$$

After integrating, we obtain:

$$\frac{u_i^{j+\frac{1}{2}}-u_i^j}{\tau}=-g\left(\frac{\zeta_{i+1}^{j+1}-\zeta_{i-1}^{j+1}}{2h_x}\right). \tag{4.2.21}$$

Expression (4.2.21) is a difference analog of equation (4.2.14). We obtained discrete analogues for equations (4.2.12), (4.2.13) and (4.2.14) on the grid (4.2.15). This difference scheme will be solved by the sweep method. For the resulting difference scheme, it is necessary to estimate the approximation error and to investigate for convergence.

4.3 The Research of a Discrete Model

Determination of the Error of Approximation

When approximating, the equations must be related to the approximate differential operator, i.e., the equations satisfy polynomials [Krylov *et al.*, 1983]. If the equations are satisfied for all polynomials of degree at most r c up to $O(h^r)$, then the difference scheme is said to have the r-th order of approximation.

For defining the approximation order the decomposition in Taylor rows of functions velocity components and level swelling concerning knot (j) and (j+1) is used. As a result, it was found that the approximation error for the problem (4.2.12), (4.2.13), and (4.2.14) is $O(\tau+h_x^2)$.

Investigation of the Stability of the Difference Scheme

The approximation condition is not sufficient for reception of the precise solution at step reduction. For such schemes maintenance of stability condition is necessary still. It is possible to present these schemes as some linear operator. It will transform function values at the moment t to function values at the moment of $t+h$. The stability condition demands, those values of this operator did not exceed in module 1+*ch*, where *c*-constant. At default of the given condition an error of the scheme quickly increases and the smaller the step, the less precise is the result. If approximation and stability conditions are satisfied the result difference schemes converges to the solving of the differential equation. According to the Courant-Friedrichs-Levy condition grid steps are limited by the following expression:

$$\frac{h_x}{\tau} < \frac{2gH}{|\tilde{u}|}. \tag{4.3.1}$$

For calculating the discrete equations (4.2.17), (4.2.19) and (4.2.21) one of economic direct methods – a *sweep* method is used [Samarsky et al, 1989]. The sweep method represents a variant of Gauss method. It is applied to special systems the linear algebraic equations with tape structure of matrix.

4.4 Results of Numerical Modeling based on Shallow Water Equations

Algorithm of the software code "BayWaves"

The software code "BayWaves" has been developed for calculation of functions velocity components and level swelling [Abbasov, Neverov, 2012]. The software code has been made and realized in language C++ in operational system Windows Vista. The algorithm of the BayWaves software code is as follows:

1. Input specification:
   - the size of the grid n;
   - step by time h_t, by space h_x;
   - the initial moment of time t;
   - the final point in time t_0;
   - the depth of the bay H;
   - setting the initial values of the velocity components u;
   - setting the initial values of the components of the elevation level ζ.

   The grid takes on values $0 \ldots N-1$, where the nodes 0 and $N-1$ are boundary, and the internal ones are $1 \ldots N-2$.
2. Calculation of velocity components in the auxiliary time layer. Depending on the grid node, the coefficients in the

central node and the coefficients for the neighboring nodes are found. In the case of a border, neighboring nodes, like the coefficients for them, exist, but one of the nodes is fictitious and does not take part in the calculation. The obtained SLAE is solved by the sweep method.

3. Calculation of the components of the function of level elevation. After the velocity components from point 2 are found, the coefficients for each node for the level elevation function are calculated. In this case, there are also dummy nodes.
4. Calculation of velocity components on the next time layer. There is a correction of the velocity component taking into account the gravity. For all nodes of the grid is expressed explicitly and identically except for the right border, where the node is not taken into account $N - 1$. Further increase the time counter by one. The software code terminates as soon as the components of the elevation function are calculated for the final time point.

Let us consider simulation features of surface gravity wave propagation under shallow-water conditions. Some hydrophysical data of shallow water area are required for simulation. We shall use the conditions of the Taganrog Bay of the Azov Sea (as it is geographically available) as a shallow water area. Though, hydrophysical conditions of other shallow water areas of World Ocean can be used.

Hydrophysical Conditions of the Azov Sea

The Azov Sea and its subsurface relief were formed at the drowning of the Azov-Kuban basin. The Azov Sea is an interior sea, its shape is comparatively simple, its coasts are relatively regular and the bottom configuration is rather simple. The Azov Sea is surrounded by escarpment and spit forms. The smooth coast grades to plane and flat bottom. The inmost depths are in the central part of the sea. Figure 4.4.1 presents a map of the depths of the Azov Sea; the inmost depth of the Azov Sea is 14m, and the mean depth is about 8m [Hydromet. Seas, 1991].

There are ten bays in the Azov Sea; three of them are of the closed type: the Taganrog Bay, the Sivash and the Taman Bay. In the northeast the strongly desalted Taganrog Bay is connected with the sea by the wide strait between the Belosarayskaya and Dolgaya Spit. The Bay's least width (about 26 km) is between the Petrushina Spit and the Tchumburskaya Spit. The Bay's bottom is intensely receding from the delta of the Don River to

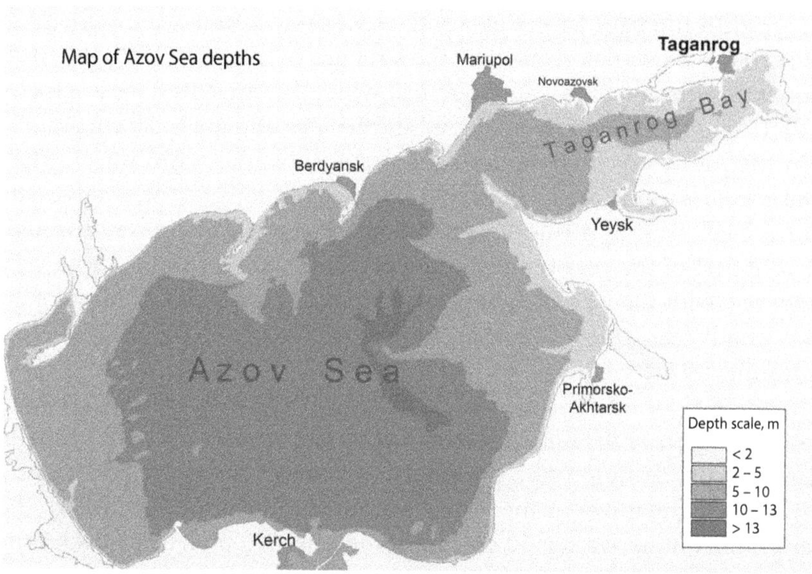

Figure 4.4.1 Map of the depths of the Azov Sea.

the sea; the average underwater gradient is 0.06%. To the West of Taganrog the Bay depth reaches 5 m, and it is less than 1 m near the Don's delta [Mamykina *et al.*, 1980].

The average annual water temperature at the sea surface is 11°C; in winter it has negative values almost throughout the entire sea. In summer, a fairly uniform surface temperature of + 24°C is established. The average salinity of the Azov Sea is 11 ‰, the water in the Taganrog Bay has a salinity of 1 ‰ near the delta of the river Don to 10 ‰ at the entrance to the bay.

The horizontal movement of water in the entire thickness of the Azov Sea is due mainly to wind. After the beginning of the wind, wind currents arise, and somewhat later and compensatory. With the cessation of the wind, the currents rapidly decay. In the autumn and winter time, the northeast and east winds prevail in the Sea of Azov with an average speed of 4-7 m/s, sometimes they reach the storm values (15 m/s and more). In the spring, the Mediterranean cyclones often pass through the sea, accompanied by southwest and west winds, at a speed of 4-6 m/s, and sometimes squalls.

The main course of the Sea of Azov forms a cycle directed counter-clockwise. Along the northern shores, the water moves westward, turning from the western coasts to the south. Near the southern coast the current goes to the east, near the Taganrog Bay it takes the western direction and closes the cycle. In the bays and inlets there are local gyrations. The velocity of the currents is also determined by the strength of the wind. With the most frequently repeated three- and four-ball wind, the currents have a velocity of 10-15 cm/s, the storm wind increases the flow velocity to 70 cm/s.

In the Azov Sea wave motions are firstly shown as wind waves. They quickly develop, and after 2 hours the wind appears, reach the steady state. On the high seas, as a rule, short and very steep waves appear. Through the cold part of the year the reigning northeast and east winds cause very rough sea, with the wave height growing to 2.1 m, even 3.0 m. More often wave length reaches 10 - 12 m and wave height is from 0.5 m up to 1 m [Shlyamin, 1977].

In the Sea of Azov, single-site seiches with a daily period are well expressed. Their nodal line runs from northwest to southeast. They usually arise after the cessation of a prolonged east wind. There are particularly noticeable fluctuations in the water level in the Taganrog Bay, where the changes are on the average 30 cm and sometimes 95 cm. Due to weak connection with the ocean, tidal fluctuations in the level in the Sea of Azov are practically absent.

The greatest changes in sea level are caused by surge-surging fluctuations in the level, which most often occur in autumn and winter. The most significant are also in the Taganrog Bay, a record magnitude of the level fluctuation is 5.8 m [Zhilyaev, 1972].

Results of Numerical Simulation

The values of the parameters of the surface gravity wave are adapted to the bathymetric conditions of the Taganrog Bay: the average depth does not exceed 5 m, the conditions of shallow water are satisfied by gravitational waves with lengths exceeding 30 m.

The length of the surface wave is l = 155.6m; depth H = 5m; velocity c = 7m/s; the wave parameter kH = 0.2 (that is, it satisfies the shallow water conditions $kH<1$). The size of the grid n = 5000; the number of nodes per wavelength is 1556, the step value over the space h_x = 0.1m.

The profile of sine surface wave in Figure 4.4.2 does not suffer special changes on its propagation way. With the growth of surface wave steepness the original sine profile is changing, primarily troughs become sloping (Figure 4.4.2*b*), further on an intermediate crest appears in the trough. On its propagation way the intermediate crest is flattening and the leading edge of the leading wave crest is steeping (Figure 4.4.2*c*).

With the less steepness the leading edge steeping is slow, i.e., it requires to accumulate nonlinear effects at long distances. The increase of wave

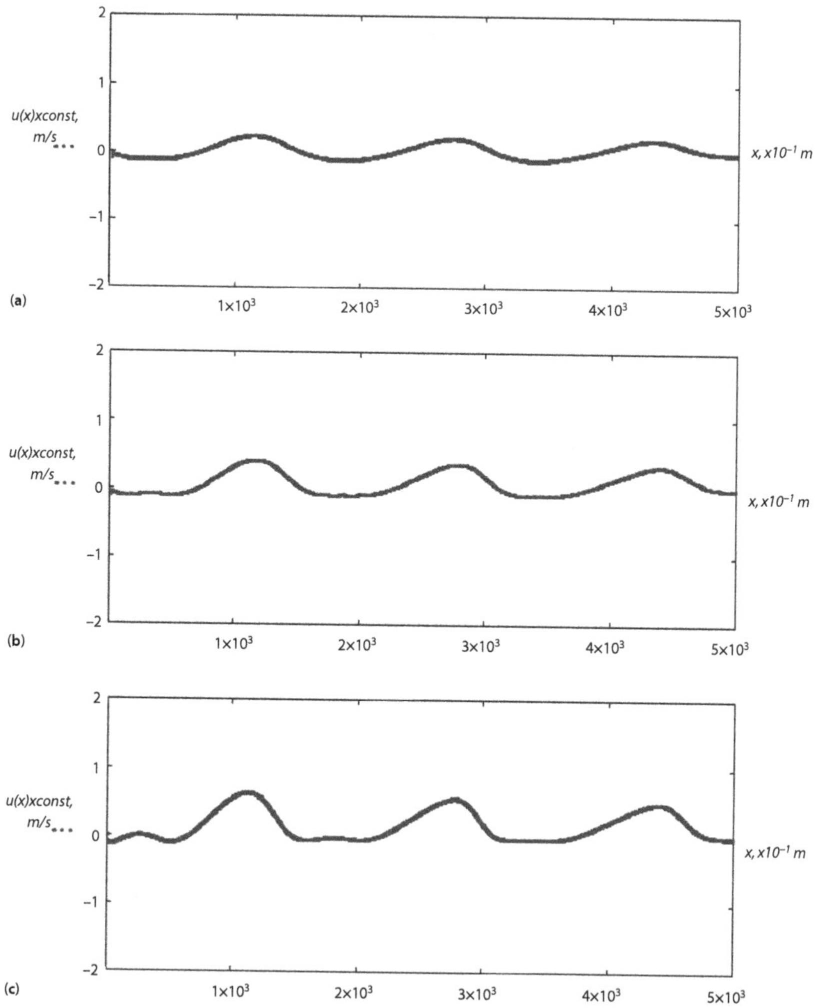

Figure 4.4.2 Dependence of the surface gravity wave profile on the initial steepness: $f = 0.045$Hz; $l = 155.6$m; H = 5m; $c = 7$m/s; $kH = 0.2$; (a) $2a/l = 0.0024$; $\varepsilon = 0.04$; (b) $2a/l = 0.0039$; $\varepsilon = 0.06$; (c) $2a/l = 0.006$; $\varepsilon = 0.09$.

steepness value results in nonlinearity increase and the surface wave comes to the leading edge steeping faster. The steeping of the surface wave leading edge is connected with the influence of nonlinear term of shallow-water equations. Approaching the land the wave crest moves faster than a trough for its bottom friction. At the moment when "the crest is catching the trough up" the leading edge becomes vertical, and the wave breaks.

Figure 4.4.3 shows the dependence of the particle velocity and the function of elevating the liquid surface for different depths. The frequency of

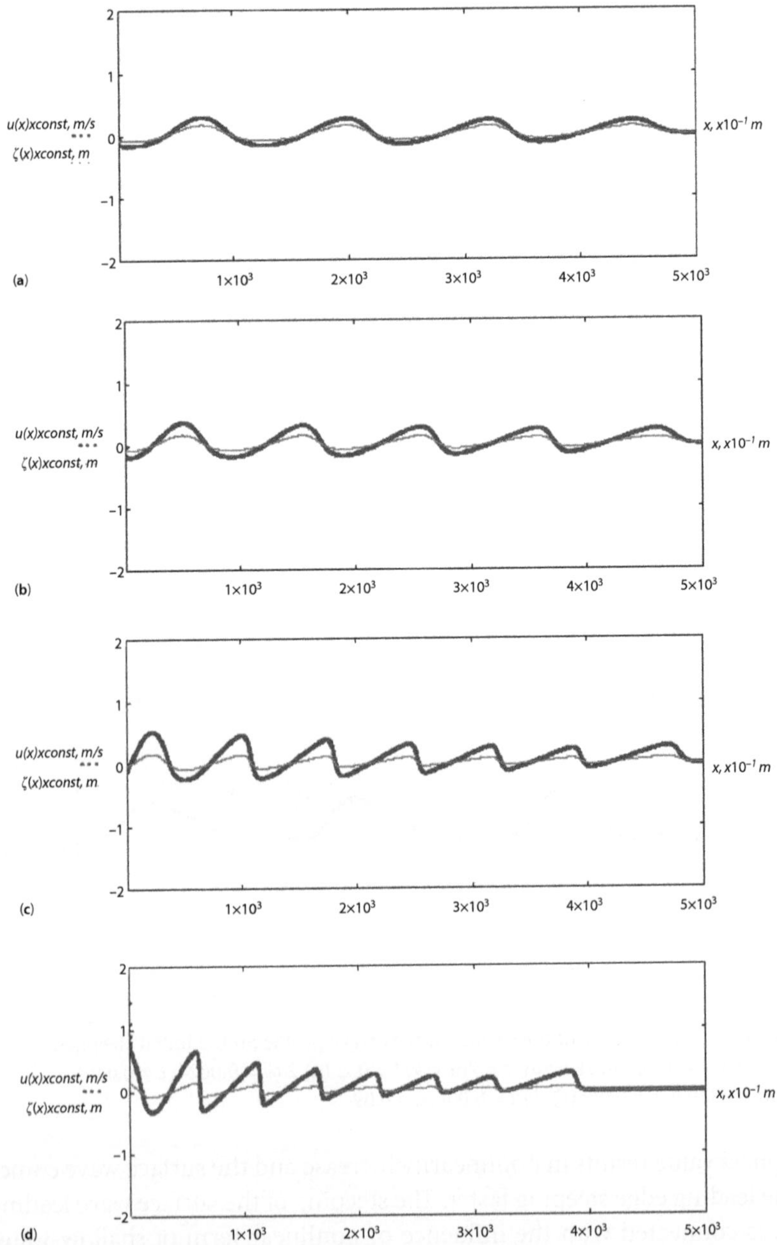

Figure 4.4.3 Dependence of the velocity of particles and the function of the elevation of the surface wave on the depth: $f = 0.045$Hz; (a) $\lambda = 120.5$m; H = 3m; $c = 5.4$m/s; $kH = 0.16$; $2a/\lambda = 0.003$; $\varepsilon = 0.06$; (b) $\lambda = 98.4$m; H = 2m; $c = 4.4$m/s; $kH = 0.13$;.$2a/\lambda = 0.0037$; $\varepsilon = 0.09$; (c) $\lambda = 69.6$м;. H = 1m; $c = 3.1$m/s; $kH = 0.09$; $2a/\lambda = 0.005$; $\varepsilon = 0.18$; (d) $\lambda = 49.2$m; H = 0,5m; $c = 2.2$m/s; $kH = 0.064$; $2a/\lambda = 0.007$; $\varepsilon = 0.36$.

the surface wave remains constant, the depth varies, the propagation velocity decreases with decreasing depth, and, consequently, the wavelength decreases. The length of the propagation path (grid size), as well as the amplitude (height) of the wave remain constant. As the depth decreases, the velocity of the particles of the medium increases, the wave contracts, and the steepness of the wave increases.

For the depth H = 3m (Figure 4.4.3a), during the time of propagation, the profile of the surface wave does not undergo special changes. For a depth of H = 1m, a sinusoidal wave approaches the steepening of the leading edge at the end of the propagation path (Figure 4.4.3c). For a depth of H = 0.5m (l = 49.2m) the surface wave does not reach the end of the propagation path, since the leading front of the wave becomes steep due to steepness and the process stops (Figure 4.4.3d). Here, the depth becomes comparable with the wave height, and the damping of the wave is also observed.

Within the observation of the shallow water conditions, let us consider changes in the profile of surface gravity waves with different frequency values. Let's select the frequencies multiples of the reduced: twice as much – f = 0.09Hz, and twice less – f = 0.0225Hz. Figure 4.4.4 and Figure 4.4.5 show

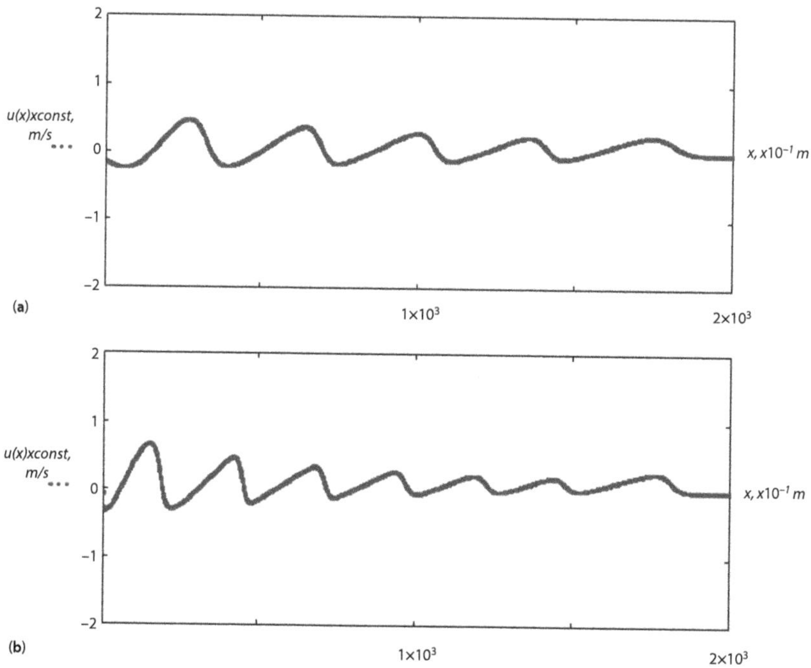

Figure 4.4.4 Dependence of the surface gravity wave profile on depth: f = 0.09Hz; (a) λ = 34.8m; c = 3.1m/s; H = 1m; $2a/\lambda$ = 0.008; kH = 0.18; ε = 0.14; (b) λ = 24.6m; c = 2.2m/s; H = 0.5m; $2a/\lambda$ = 0.01; kH = 0.13; ε = 0.2.

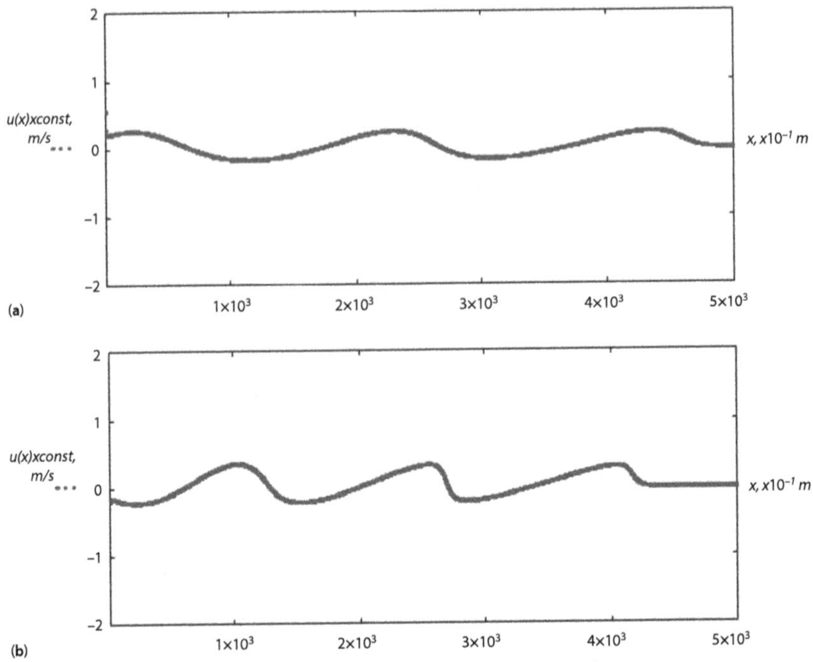

Figure 4.4.5 Dependence of the surface gravity wave profile on depth: $f = 0.0225$Hz; (a) $l = 196.8$m; $c = 4.4$m/s; H = 2m; $2a/l = 0.0015$; $kH = 0.06$; $\varepsilon = 0.08$; (b) $l = 139.2$m; $c = 3.13$m/s; H = 1m; $2a/l = 0.008$; $kH = 0.04$; $\varepsilon = 0.12$.

surface wave profiles with different frequencies for different depths. For these dependencies, the described regularities are preserved. At a constant amplitude, a decrease in depth leads to an increase in the steepness, and, consequently, to the steepening of the leading edge of the wave.

Modeling Refraction of Surface Wave

Let's proceed to numerical modeling of the refraction of nonlinear gravity surface waves in the coastal zone. To do this, we simulate coastal slopes of different steepness in the conditions of the bay. The bottom lines are shown in Figure 4.4.6 (the depth scale is increased). They are modeled on the graphs of power functions. The depth decreases from H = 3m to zero, the slope steepness does not exceed 0.01^0.

$$s3(x) = 2{,}7\cdot\left(\left(\frac{x}{n}\right)+\left(\frac{x}{n}\right)^2-\left(\frac{x}{n}\right)^3\right);s4(x) = 2{,}7\cdot\left(\left(\frac{x}{n}\right)+9\cdot\left(\frac{x}{n}\right)^2\right).$$

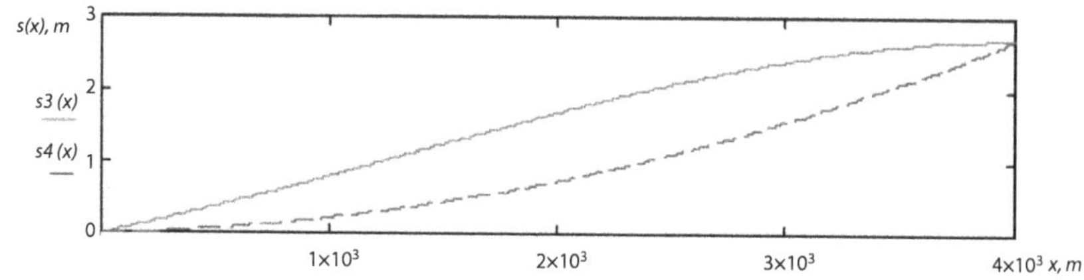

Figure 4.4.6 Lines of the bottom according to analytic expressions: where n is the size of the grid:

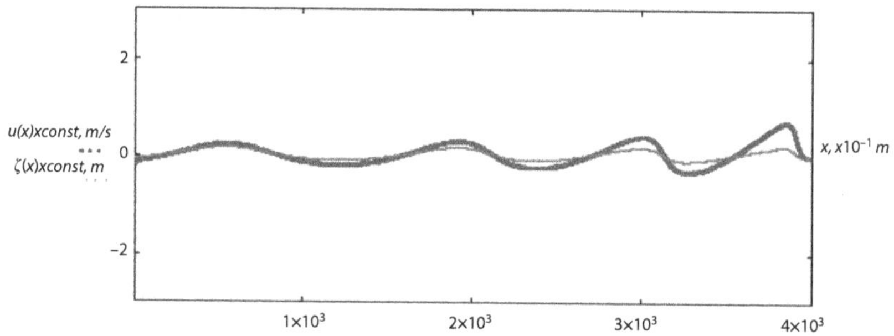

Figure 4.4.7 Transformation of the profile of the surface gravity wave as it approaches the shore, initial wave parameters: $f = 0.045$Hz; $l = 155.6$m; H = 5m; $c = 7$m/s; $kH = 0.2$; $2a/l = 0.0023$; $\varepsilon = 0.036$, the bottom rises linearly; mesh size $n = 4000$.

In this case, the depth values in Figure 4.4.6 are taken less than the average depth of the Bay to accelerate the refraction process. The bottom line $s3(x)$ is more shallow, and the line $s4(x)$ is steeper along the shore. In the beginning, we will use a uniformly (linearly) rising bottom line for the depth of the bay H = 5м.

Figure 4.4.7 shows the dependencies for the particle velocity and the surface wave surfacing function when approaching the shore. The initial surface wave is sinusoidal, with a decrease in depth, the propagation velocity decreases. This leads to a decrease in the wavelength with constant frequency and observance of the law of conservation of energy.

Consequently, the profile of the wave is compressed and the velocity of the particles increases as in Figure 4.4.7. Along with the decrease in the wavelength at the shore, we observe the steepening of the leading edge of the crest of the wave, which in future will lead to its breaking in the surf zone.

Figure 4.4.8 shows the dependencies for the particle velocity of the medium when approaching the shore of the surface wave along the bottom lines with different steepness. Figure 4.4.8a shows the profile of the surface

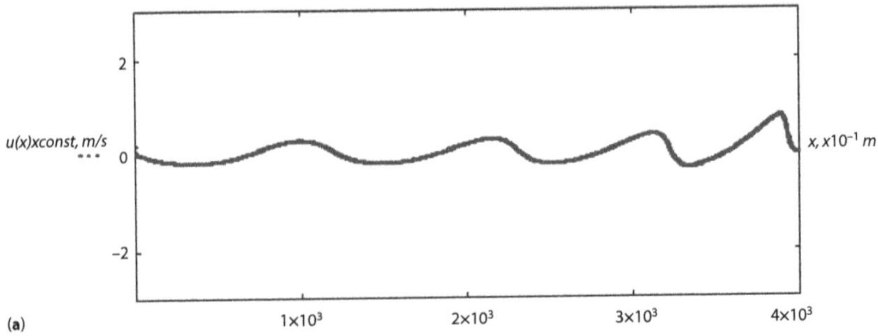

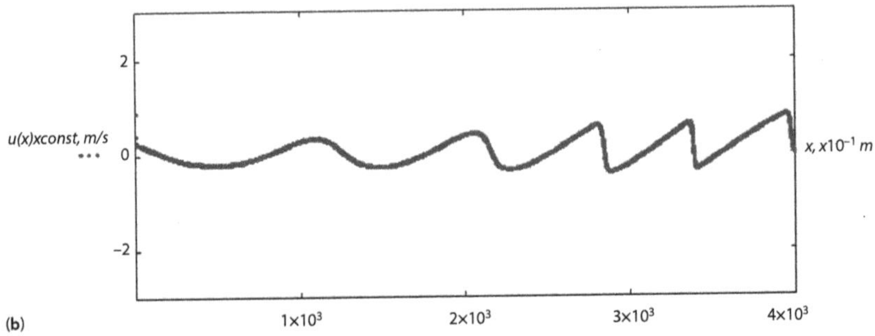

Figure 4.4.8 Transformation of the profile of the surface gravity wave as it approaches the shore, initial wave parameters: $f = 0.045$Hz; $\lambda = 120.5$m; H = 3m; $c = 5.4$m/s; $kH = 0.16$; $2a/\lambda = 0.003$; $\varepsilon = 0.06$; (a) steepened bottom line - $s4(x)$; (b) flat bottom line - $s3(x)$.

wave propagating along the bottom with a steep rise at the shore $s4(x)$. Figure 4.4.8b shows the wave profile for the bottom with a gentle rise near the shore $s3(x)$.

From the comparison of profiles, it can be noted that, with a shallow coastline, the surface wave is more quickly steepened, and, therefore, can break without reaching the shore. For a slowly rising bottom, the wave front is steepened, approaching the shoreline.

4.5 Discussion and Comparison of Results

To verify the adequacy of our results of numerical modeling of the propagation of nonlinear surface gravity waves in shallow water, we compare them with existing numerical and experimental data.

Figure 4.5.1a shows a numerical model of distortion of the profile of a nonlinear surface wave [Shimozono, Sato, 2009]. Distortions of the profile are modeled on the basis of the Boussinesq equation. For comparison, Figure 4.5.1b shows the surface wave profiles obtained by us on the basis

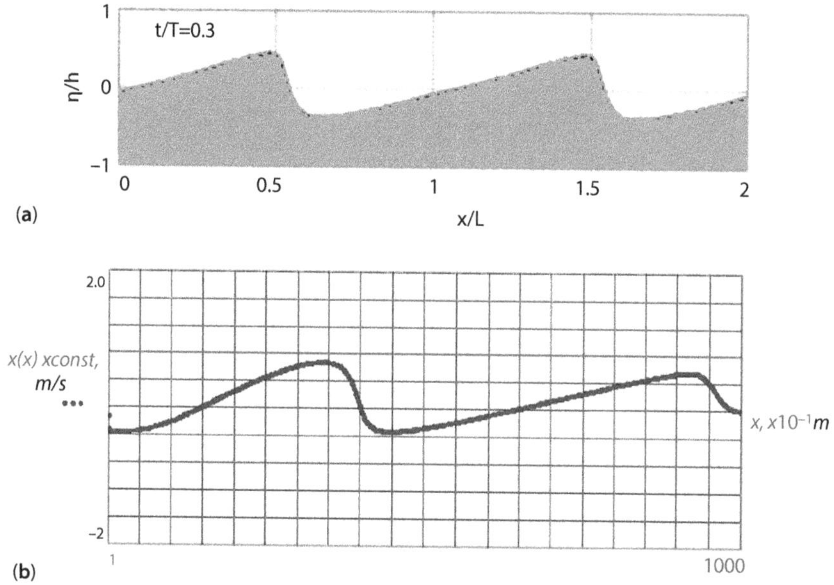

Figure 4.5.1 Gradual distortion of the profile of the surface gravity wave (a): $\varepsilon = 0.75$; layer $t/T = 0.3$– nonlinear dispersion model [Shimozono, Sato, 2009]; (b): $f = 0.045$Hz; $\lambda = 49.2$m; H = 0.5m; $c = 2.2$m/s; $\varepsilon = 0.4$; $t = 39.9$s – model of nonlinear shallow water equations.

of a numerical solution of the shallow water equations. As a result of the comparison, we can note:

- The wave with the sinusoidal profile begins to distort, the front slope of the ridge becomes steep, the back slope is stretched and becomes flat. These distortions coincide both for the model in Figure 4.5.1a, and for our model in Figure 4.5.1b.

The paper [Kawasaki, 1999] is devoted to numerical and experimental studies of the influence of a flooded breakwater on the propagation of the surface wave (the length of the laboratory basin is 30m, the width is 0.7m, the depth is 0.95m, the wave periods are T = 0.8, 1.2, 1.68s) A two-dimensional finite-difference numerical model is created on the basis of the Navier-Stokes equations. To test the model, the results of laboratory experiments are presented.

Figure 4.5.2 shows the results of numerical simulation and experimental profiles of the surface wave at different stages of propagation. Figure 4.5.3 shows the results of a numerical calculation of surface wave profiles with

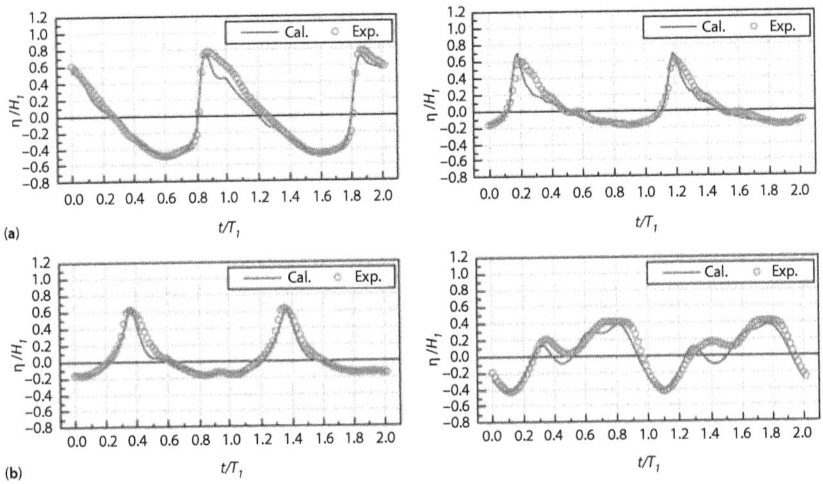

Figure 4.5.2 Calculated and experimental surface wave profiles (shore on the left): ($2a/\lambda = 0.03$, $h/\lambda = 0.2$), (a) at the beginning of the wave breaker ($x/\lambda = 0{,}1$); (b) above the wavebreaker ($x/\lambda = 0{,}3$); (c) above the wavebreaker ($x/\lambda = 0{,}4$); *d*) behind the breakwater ($x/\lambda = 0{,}7$).

different initial steepness values based on our model. On the basis of their comparison, we can note the following:

- the experimental profiles of surface waves in front of the breakwaters begin to be distorted due to a decrease in depth, the leading edge of the crest begins to become steep, the troughs are smoothed (Figure 4.5.2a, b);
- with further spread over the breakwater, crests are sharpened, an intermediate crest appears in the hollow;
- in Figure 4.5.3a, the surface wave due to the effect of depth on further propagation from the sinusoidal is transformed into a nonlinear one with a steep leading edge, as in the initial stage of the experimental path in Figure 4.5.2a;
- an increase in the initial steepness for our model leads to the appearance of an intermediate crest in the trough (Figure 4.5.3b), similar to the experimental profile in Figure 4.5.2c;
- After the breakwater, the depth increases, and because of the dispersion of the crests, the waves are scattered. This stage of transformation of the wave profile is no longer described by the shallow water equations.

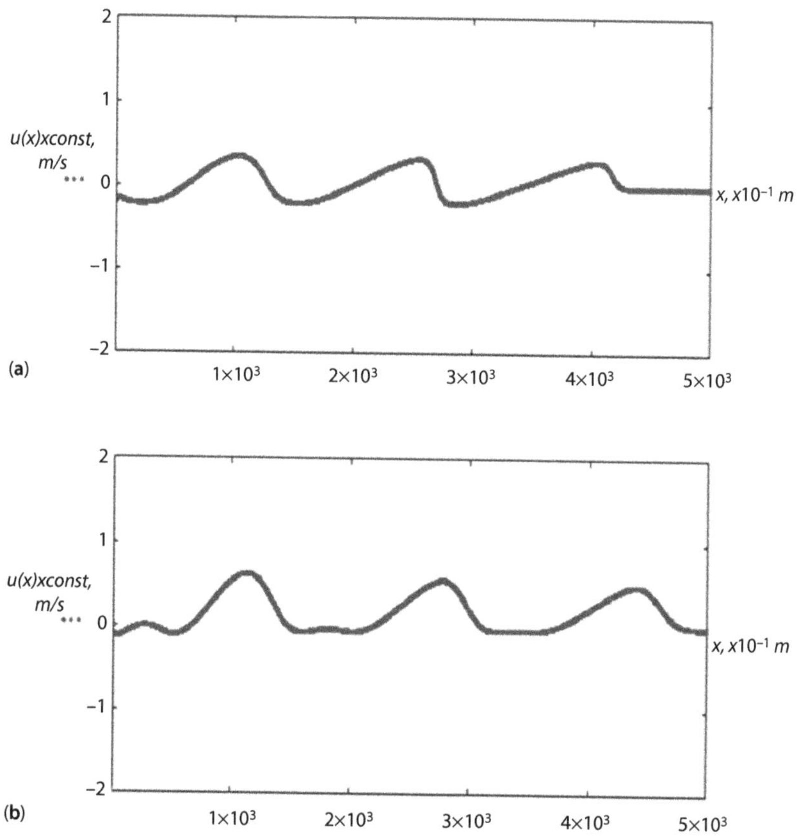

Figure 4.5.3 Results of numerical calculation of surface wave profiles with different initial steepness values: (a) $f = 0.0225$Hz; $l = 139.2$m; H = 1m; $2a/l = 0.008$; $\varepsilon = 0.12$; (b) $f = 0.045$ Hz; $l = 155.6$m; H = 5m; $2a/l = 0.006$; $\varepsilon = 0.09$.

Consequently, the results of the numerical simulation of nonlinear surface gravity waves on the basis of shallow water equations as a whole are in good agreement with experimental measurements.

After comparing the results of the studies with experimental data, to confirm the correctness of the proposed analytical method [Abbasov, 2009; Doklady], we compare its results with the numerical calculation.

The following Figure 4.5.4 and Figure 4.5.5 show the particle velocities of the same surface gravity wave calculated within our models (in two stages of propagation):

- *semigraphical method based on the nonlinear equation;*
- *numerical calculation based on shallow water equations.*

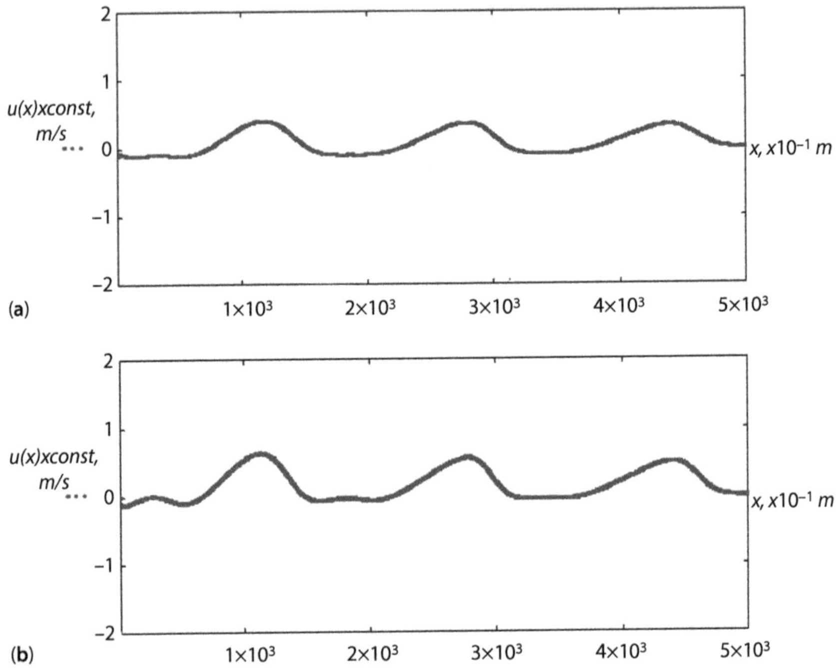

Figure 4.5.4 Surface wave profiles – numerical calculation.

From a comparison of surface wave profiles, it can be noted:

- the initial stage of the sharpening of the crests of surface waves is well described both by the graphoanalytical method (Figure 4.5.5a) and by numerical calculation (Figure 4.5.4a, its steepness is $2a/\lambda = 0.004$);
- the stage of the surface wave steepening is also well described both by the semigraphical method (Figure 4.5.5b) and by numerical calculation (Figure 4.5.4b).

Therefore, it can be noted that the proposed semigraphical method is completely correct for describing nonlinear surface waves in shallow water.

Conclusions

In the fourth chapter the questions of modeling the distribution of surface gravity waves based on computational solution of the equation of

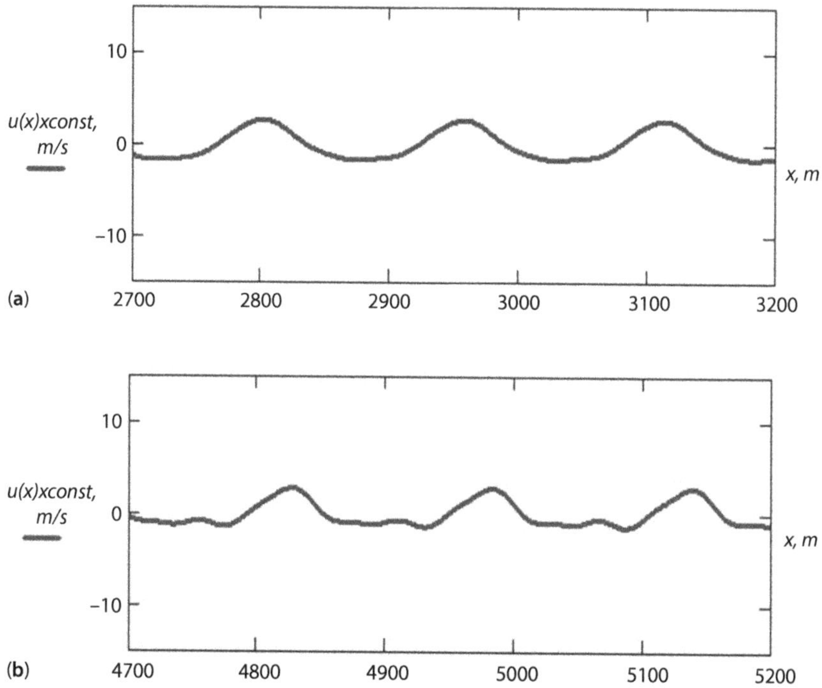

Figure 4.5.5 Surface wave profiles are the semigraphical method, the main parameters of the surface gravity wave: $f = 0.045$Hz; $l = 155.6$m; H = 5m; $c = 7$m/s; $kH = 0.2$; $2a/l = 0.006$; $\varepsilon = 0.1$.

shallow water are examined. To assess the status of the issue the studies on computational modeling of wave hydrodynamic tasks were analyzed. The works of the computational solution, both nonlinear and nonlinear-dispersion equations of shallow water, and nonlinear-dispersion equations of Boussinesq, Korteweg-de Vries were described.

An analysis of the modeling of wave processes in the incompressible fluid, based on the Navier-Stokes equations, was carried out. The computational methods for solving the Navier-Stokes equations for the incompressible fluid have been considered. The works of the computational modeling of surface waves based on the nonlinear equations of Navier-Stokes were presented. The 2D and 3D modeling of run-up of the surface waves for shallow water to the coastal formations were noted.

The objective, the initial and the boundary conditions were formulated. After using the physical process split method, a system of three differential equations was obtained. A differential scheme was used for computational solution of the differential equations. The differential scheme was built using a integral-interpolation method on a uniform grid in an implicit scheme.

Further, a discrete analogy of the system of equations was created, the order of approximation was determined, and the conditions of sustainability of the discrete model were examined. The sweep method was used to calculate the system of equations. A "BayWaves" program was developed to calculate particle velocity and a free surface elevation function, and a description of the structure and algorithm of the program were presented.

The results of the numerical calculation of particle velocity and the function of the elevation of the free surface of the gravity wave for the different values of the initial steepness and the shallow water depth values were provided. For the computational modeling of the refraction of nonlinear surface gravity waves at the coastal region the beach approach of various slopes under the conditions of creek are modeled. Graphs for particle velocity and the surface wave elevation function were presented.

In order to verify the adequacy of the results of the computational modeling, comparisons were made with existing numerical and experimental data. In order to verify the correctness of the previously proposed analytical expressions, the results of the modeling were also compared to the graphic method with the results of the computational calculation. As the result of the comparison computation confirmed the propriety of the proposed semigraphical method based on the comparison of the surface gravity wave profiles with the same properties.

The results of the studies in this chapter have been published in the following main works: [Abbasov, 2010 Fizmatlit], [Abbasov, 2010], [Abbasov, 2012], [Abbasov, Neverov, 2012].

5

Two-Dimensional Numerical Simulation of the Run-Up of Nonlinear Surface Gravity Waves

5.1 Statement of the Problem

Boundary and initial conditions

To describe the two-dimensional surface gravity waves on the fluid surface with viscosity, the two-dimensional Navier–Stokes equation, the continuity equation for incompressible fluid, and the equation of hydrodynamic pressure are used [Fletcher, 1991], [Samarsky, 1989]:
the Navier–Stokes equations:

$$\frac{\partial u}{\partial t}+u\frac{\partial u}{\partial x}+w\frac{\partial u}{\partial z}=-\frac{1}{\rho}\frac{\partial P}{\partial x}+\mu\frac{\partial^2 u}{\partial x^2}+\eta\frac{\partial^2 u}{\partial z^2}, \qquad (5.1.1)$$

$$\frac{\partial w}{\partial t}+u\frac{\partial w}{\partial x}+w\frac{\partial w}{\partial z}=-\frac{1}{\rho}\frac{\partial P}{\partial z}+\mu\frac{\partial^2 w}{\partial x^2}+\eta\frac{\partial^2 w}{\partial z^2}+g, \qquad (5.1.2)$$

the continuity equation for incompressible fluid:

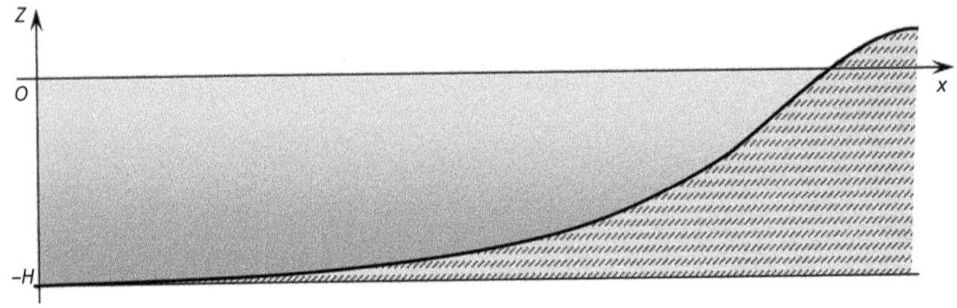

Figure 5.1.1 Geometry of the problem.

$$\frac{\partial u}{\partial x}+\frac{\partial w}{\partial z}=0, \tag{5.1.3}$$

the full hydrodynamic pressure with account of depth is determined by:

$$P(x,z,t)=p(x,z,t)+\rho gz, \tag{5.1.4}$$

where u and w are the horizontal and vertical components of the water particle motion velocity vector $V(u,w)$, ρ is the fluid density, g is the gravitational constant, and μ and η are the horizontal and vertical components of the turbulent viscosity coefficient.

According to the geometry of the problem, the Ox-axis of the coordinate system is aligned with the undisturbed fluid surface and is directed toward the coast; the Oz-axis is directed vertically upward (Figure 5.1.1).

Boundary Conditions

At the bottom of area conditions impermeable and slidings are assumed:

$$\begin{gathered}V_n(x,z,t)=0, \frac{\partial}{\partial n}P(x,z,t)=0,\\ \frac{\partial}{\partial x}u(x,z,t)=0, \frac{\partial}{\partial z}w(x,z,t)=0,\\ \rho\eta\frac{\partial}{\partial z}u(x,z,t)=-\tau_x(t), \rho\mu\frac{\partial}{\partial x}w(x,z,t)=-\tau_z(t);\end{gathered} \tag{5.1.5}$$

- on the free surface of a fluid – the dynamic and kinematic condition:

$$\frac{\partial}{\partial n}u(x,z,t)=0,\frac{\partial}{\partial n}w(x,z,t)=0,$$

$$w(x,z,t)=\frac{1}{g\rho}\frac{\partial P}{\partial t},\frac{\partial}{\partial n}P(x,z,t)=0; \qquad (5.1.6)$$

- on the left lateral border (there is a source), the slidings condition:

$$\frac{\partial}{\partial n}u(x,z,t)=0,\frac{\partial}{\partial n}w(x,z,t)=0,\frac{\partial}{\partial n}P(x,z,t)=\alpha, \qquad (5.1.7)$$

- the right lateral border is continuation of a bottom, conditions are similar;

where $\tau_x(t)$, $\tau_z(t)$ – are the tangential stress components at the fluid bottom, and a – is a parameter specified from initial conditions.

The tangential stress components are defined by following expressions (according to the Van Dorn law [Fletcher, 1991], [Roach, 1980]):

$$\tau_x(t)=\rho C_p(|V|)u|V|,\quad \tau_z(t)=\rho C_p(|V|)w|V|, \qquad (5.1.8)$$

where $C_p(|V|)=\begin{cases}0.0088; & |V|<6.6m/s,\\ 0.0026; & |V|\geq 6.6m/s\end{cases}$ – dimensionless coefficient.

The horizontal and vertical components of the turbulent viscosity coefficient accept following values μ,η ($0<\mu<1m^2/s$, $0<\eta<1m^2/s$).

At the initial time, the fluid is still, and following conditions are satisfied:

$$t=0:\ P(x,z,0)=\rho gz,\ u(x,z,0)=0,\ w(x,z,0)=0.$$

Splitting of Equations

We approximate equations (5.1.1) and (5.1.2) first with respect to the time variable and introduce an intermediate layer [Harlow, Welch, 1965], [Harlow, 1967]:

$$\frac{u^{n+1}-u^{n+\sigma}+u^{n+\sigma}-u^{n}}{\tau}+u\frac{\partial u}{\partial x}+w\frac{\partial u}{\partial z}=-\frac{1}{\rho}\frac{\partial P}{\partial x}+\mu\frac{\partial^2 u}{\partial x^2}+\eta\frac{\partial^2 u}{\partial z^2}, \qquad (5.1.9)$$

$$\frac{w^{n+1}-w^{n+\sigma}+w^{n+\sigma}-w^{n}}{\tau}+u\frac{\partial w}{\partial x}+w\frac{\partial w}{\partial z}$$
$$=-\frac{1}{\rho}\frac{\partial P}{\partial z}+\mu\frac{\partial^2 w}{\partial x^2}+\eta\frac{\partial^2 w}{\partial z^2}+g. \tag{5.1.10}$$

Apply the method of *splitting* according to physical processes to equations (5.1.1) and (5.1.2)

$$\frac{u^{n+\sigma}-u^{n}}{\tau}+u\frac{\partial u}{\partial x}+w\frac{\partial u}{\partial z}=\mu\frac{\partial^2 u}{\partial x^2}+\eta\frac{\partial^2 u}{\partial z^2}, \tag{5.1.11}$$

$$\frac{w^{n+\sigma}-w^{n}}{\tau}+u\frac{\partial w}{\partial x}+w\frac{\partial w}{\partial z}=\mu\frac{\partial^2 w}{\partial x^2}+\eta\frac{\partial^2 w}{\partial z^2}+g, \tag{5.1.12}$$

$$\frac{u^{n+1}-u^{n+\sigma}}{\tau}=-\frac{1}{\rho}\frac{\partial P}{\partial x}, \tag{5.1.13}$$

$$\frac{w^{n+1}-w^{n+\sigma}}{\tau}=-\frac{1}{\rho}\frac{\partial P}{\partial z}. \tag{5.1.14}$$

Perform differentiation with respect to the coordinates x and y for the equations (5.1.13), (5.1.14) and sum the result obtained (taking into account the continuity equation):

$$\frac{\partial^2 P}{\partial x^2}+\frac{\partial^2 P}{\partial z^2}=\frac{\rho}{\tau}\left(\frac{\partial u^{n+\sigma}}{\partial x}+\frac{\partial w^{n+\sigma}}{\partial z}\right). \tag{5.1.15}$$

According to the *splitting* method, the problem under investigation will consist of three subtasks. The *first subtask* includes the system of equations (5.1.11) and (5.1.12) for calculating the velocity field without taking the pressure into account:

The boundary conditions in this case are:

- at the bottom of the region:

$$\frac{\partial}{\partial x}u(x,z,t)=0,\frac{\partial}{\partial z}w(x,z,t)=0, \tag{5.1.16}$$

$$\rho\eta\frac{\partial}{\partial z}u(x,z,t)=-\tau_x(t),\rho\eta\frac{\partial}{\partial x}w(x,z,t)=-\tau_z(t);$$

- on the free surface of the liquid and the lateral boundary:

$$\frac{\partial}{\partial n}u(x,z,t)=0,\frac{\partial}{\partial n}w(x,z,t)=0. \tag{5.1.17}$$

The *second subtask* includes equation (5.1.15) for calculating pressure without taking into account the influence of gravity.

The boundary conditions in this case are:

- at the bottom of the region:

$$V_n(x,z,t)=0,\frac{\partial}{\partial n}P(x,z,t)=0,\frac{\partial}{\partial x}u(x,z,t)$$
$$=0,\frac{\partial}{\partial z}w(x,z,t)=0; \tag{5.1.18}$$

- on the free surface of the liquid:

$$\frac{\partial}{\partial n}u(x,z,t)=0,\frac{\partial}{\partial n}w(x,z,t)=0,w(x,z,t)$$
$$=\frac{1}{g\rho}\frac{\partial P}{\partial t},\frac{\partial}{\partial n}P(x,z,t)=0; \tag{5.1.19}$$

- at the lateral border:

$$\frac{\partial}{\partial n}u(x,z,t)=0,\frac{\partial}{\partial n}w(x,z,t)=0,V_n(x,z,t)$$
$$=0,\frac{\partial}{\partial n}P(x,z,t)=\alpha. \tag{5.1.20}$$

The third subtask consists of the system of equations (5.1.13) and (5.1.14) for calculating the velocity field taking into account the effect of pressure.

The boundary conditions in this case are:

$$\frac{\partial}{\partial n}P(x,z,t)=0. \tag{5.1.21}$$

We approximate the equations (5.1.11) ÷ (5.1.15) with respect to the space variables. For this we use the integro-interpolation method.

5.2 Construction of a Discrete Finite-Volume Model

The computational domain in the spatial directions is a rectangle. For the numerical realization of a discrete mathematical model of the problem of wave hydrodynamics, we introduce a uniform grid [Samarsky, 1989]:

$$\varpi_h = \begin{Bmatrix} t^n = n\tau,\ x_i = ih_x,\ z_k = kh_z;\ i = \overline{0..N_x},\ k = \overline{0..N_z}; \\ N_t\tau = T,\ N_x h_x = l_x,\ N_z h_z = l_z \end{Bmatrix},$$

where t– a step in time, h_x, h_z– steps in space, N_t– the upper border in time, N_x,N_z– borders across space.

The calculated cells are rectangles that can be completely filled, partially filled, or empty. The cell centers and nodes are spaced apart $h_x/2$ and $h_z/2$ on the corresponding coordinates x, z. Denote the completeness of the cell through [Sukhinov, Chistyakov, 2011].

In this case, the velocity field and pressure are calculated at the vertices of this cell: (i,k), $(i-1,k)$, $(i,k-1)$, $(i-1,k-1)$ (Figure 5.2.1). The degree of cell occupancy is determined by the pressure of the liquid column inside the cell. If the average pressure in the nodes of the cell is greater than the pressure of the liquid column inside the cell, the cell is considered to be completely filled $o_{i,k} = 1$.

In the general case, the cell occupancy can be calculated from the following expression:

$$o_{i,k} = \frac{P_{i,k}H(P_{i,k}) + P_{i-1,k}H(P_{i,k}) + P_{i,k-1}H(P_{i,k}) + P_{i+1,k+1}H(P_{i,k})}{2\rho g h_z}, \quad (5.2.1)$$

where $H(x) = \begin{cases} 1, & x \geq 0, \\ 0, & x < 0 \end{cases}$ - Heaviside function.

In the neighborhood of the node (i,k) there are cells (i,k), $(i+1,k)$, $(i,k+1)$, $(i+1,k+1)$, as shown in Figure 5.2.2.

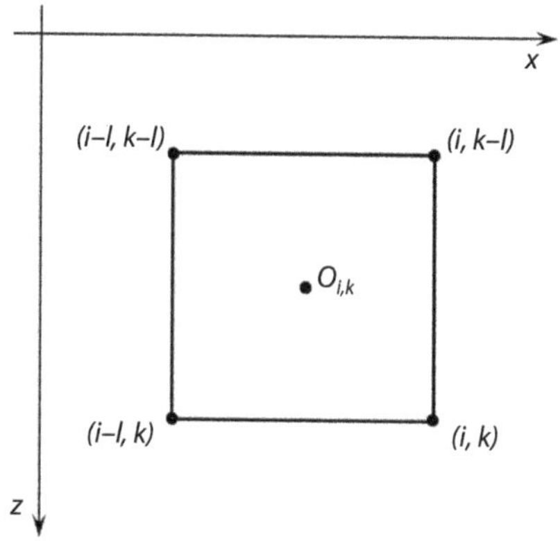

Figure 5.2.1 Location of the cell with adjacent nodes.

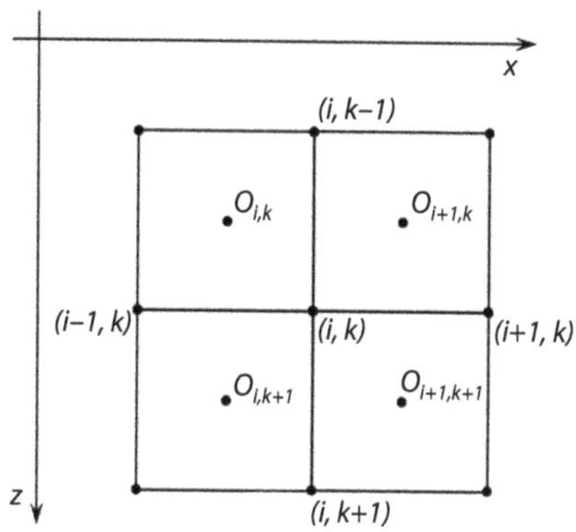

Figure 5.2.2 Location of nodes relative to cells.

To describe filling of regions in a neighborhood of a cell, we introduce the following coefficients: k_0, k_1, k_2, k_3, k_4. The value of the coefficients characterizes the filling of the regions Θ_m:

$$\begin{aligned}
&k_0 - \Theta_0 : x \in (x_{i-1}, x_{i+1}), \quad z \in (z_{k-1}, z_{k+1});\\
&k_1 - \Theta_1 : x \in (x_i, x_{i+1}), \quad z \in (z_{k-1}, z_{k+1});\\
&k_2 - \Theta_2 : x \in (x_{i-1}, x_i), \quad z \in (z_{k-1}, z_{k+1});\\
&k_3 - \Theta_3 : x \in (x_{i-1}, x_{i+1}), \quad z \in (z_k, z_{k+1});\\
&k_4 - \Theta_4 : x \in (x_{i-1}, x_{i+1}), \quad z \in (z_{k-1}, z_k);
\end{aligned} \tag{5.2.2}$$

Denote by D_m the filled parts of the regions Θ_m, where $m=\overline{0..4}$. Then the coefficients k_m can be calculated by the following expressions:

$$(k_m)_{i,k}=\frac{S_{D_m}}{S_{\Theta_m}},(k_0)_{i,k}=\frac{o_{i,k}+o_{i+1,k}+o_{i+1,k+1}+o_{i,k+1}}{4},$$
$$(k_1)_{i,k}=\frac{o_{i+1,k}+o_{i+1,k+1}}{2},(k_2)_{i,k}=\frac{o_{i,k}+o_{i,k+1}}{2}, \quad (5.2.3)$$
$$(k_3)_{i,j}=\frac{o_{i+1,k+1}+o_{i,k+1}}{2},(k_4)_{i,j}=\frac{o_{i,k}+o_{i+1,k}}{2},$$

For the first subtask of the system of equations (5.1.11), (5.1.12), the boundary conditions generally take the following form:

$$\frac{\partial}{\partial x}u(x,z,t)=\alpha_{u,x}u+\beta_{u,x},\frac{\partial}{\partial x}w(x,z,t)=\alpha_{w,x}w+\beta_{w,x}, \quad (5.2.4)$$
$$\frac{\partial}{\partial z}u(x,z,t)=\alpha_{u,z}u+\beta_{u,z},\frac{\partial}{\partial z}w(x,z,t)=\alpha_{w,z}w+\beta_{w,z}.$$

Integrate over the region D_0 the equation (5.1.11):

$$\iint_{D_0}\left(\frac{u^{n+\sigma}-u^n}{\tau}+u\frac{\partial u}{\partial x}+w\frac{\partial u}{\partial z}\right)dxdz=\iint_{D_0}\left(\mu\frac{\partial^2 u}{\partial x^2}+\eta\frac{\partial^2 u}{\partial z^2}\right)dxdz. \quad (5.2.5)$$

As a result of final integration, we can obtain the following expression:

$$(k_0)_{i,k}\frac{u_{i,k}^{n+\sigma}-u_{i,k}^n}{\tau}+(k_1)_{i,k}u_{i+1/2,k}\frac{u_{i+1,k}-u_{i,k}}{2h_x}+(k_2)_{i,k}u_{i-1/2,k}\frac{u_{i,k}-u_{i-1,k}}{2h_x}+$$
$$+(k_3)_{i,k}w_{i,k+1/2}\frac{u_{i,k+1}-u_{i,k}}{2h_z}+(k_4)_{i,k}w_{i,k-1/2}\frac{u_{i,k}-u_{i,k-1}}{2h_z}=$$
$$=(k_1)_{i,k}\mu_{i+1/2,k}\frac{u_{i+1,k}-u_{i,k}}{h_x^2}-(k_2)_{i,k}\mu_{i-1/2,k}\frac{u_{i,k}-u_{i-1,k}}{h_x^2}-$$
$$-\left((k_1)_{i,k}-(k_2)_{i,k}\right)\mu_{i,k}\frac{\alpha_{u,x}u_{i,k}+\beta_{u,x}}{h_x}+$$
$$+(k_3)_{i,k}\eta_{i,k+1/2}\frac{u_{i,k+1}-u_{i,k}}{h_z^2}-(k_4)_{i,k}\eta_{i,k-1/2}\frac{u_{i,k}-u_{i,k-1}}{h_z^2}-$$

$$-\left((k_3)_{i,k}-(k_4)_{i,k}\right)\eta_{i,k}\frac{\alpha_{u,z}u_{i,k}+\beta_{u,z}}{h_z}. \tag{5.2.6}$$

Similarly, we can obtain an approximation of the second equation (4.2.4) of the first subtask with boundary conditions:

$$(k_0)_{i,k}\frac{w_{i,k}^{n+\sigma}-w_{i,k}^{n}}{\tau}+(k_1)_{i,k}u_{i+1/2,k}\frac{w_{i+1,k}-w_{i,k}}{2h_x}+(k_2)_{i,k}u_{i-1/2,k}\frac{w_{i,k}-w_{i-1,k}}{2h_x}+$$

$$+(k_3)_{i,k}w_{i,k+1/2}\frac{w_{i,k+1}-w_{i,k}}{2h_z}+(k_4)_{i,k}w_{i,k-1/2}\frac{w_{i,k}-w_{i,k-1}}{2h_z}=$$

$$=(k_1)_{i,k}\mu_{i+1/2,k}\frac{w_{i+1,k}-w_{i,k}}{h_x^2}-(k_2)_{i,k}\mu_{i-1/2,k}\frac{w_{i,k}-w_{i-1,k}}{h_x^2}-$$

$$-\left((k_1)_{i,k}-(k_2)_{i,k}\right)\mu_{i,k}\frac{\alpha_{w,x}w_{i,k}+\beta_{w,x}}{h_x}+$$

$$+(k_3)_{i,k}\eta_{i,k+1/2}\frac{w_{i,k+1}-w_{i,k}}{h_z^2}-(k_4)_{i,k}\eta_{i,k-1/2}\frac{w_{i,k}-w_{i,k-1}}{h_z^2}-$$

$$-\left((k_3)_{i,k}-(k_4)_{i,k}\right)\eta_{i,k}\frac{\alpha_{w,z}w_{i,k}+\beta_{w,z}}{h_z}+(k_0)_{i,k}g. \tag{5.2.7}$$

In the case of internal nodes and boundary nodes located on the surface of the reservoir and on the lateral boundary:

$$\alpha_{u,x}=0,\quad \alpha_{w,z}=0,\quad \beta_{u,x}=0,\quad \beta_{w,z}=0. \tag{5.2.8}$$

For boundary nodes located at the bottom of the region under consideration and describing the motion of the fluid, equation (5.2.6) takes the form:

$$(k_0)_{i,k}\frac{u_{i,k}^{n+\sigma}-u_{i,k}^{n}}{\tau}+(k_1)_{i,k}u_{i+1/2,k}^{n}\frac{u_{i+1,k}^{n+\sigma/2}-u_{i,k}^{n+\sigma/2}}{2h_x}+$$

$$+(k_2)_{i,k}u_{i-1/2,k}^{n}\frac{u_{i,k}^{n+\sigma/2}-u_{i-1,k}^{n+\sigma/2}}{2h_x}+(k_3)_{i,k}w_{i,k+1/2}^{n}\frac{u_{i,k+1}^{n+\sigma/2}-u_{i,k}^{n+\sigma/2}}{2h_z}+$$

$$+(k_4)_{i,k} w^n_{i,k-1/2} \frac{u^{n+\sigma/2}_{i,k} - u^{n+\sigma/2}_{i,k-1}}{2h_z} = (k_1)_{i,k} \mu_{i+1/2,k} \frac{u^{n+\sigma/2}_{i+1,k} - u^{n+\sigma/2}_{i,k}}{h_x^2} -$$

$$-(k_2)_{i,k} \mu_{i-1/2,k} \frac{u^{n+\sigma/2}_{i,k} - u^{n+\sigma/2}_{i-1,k}}{h_x^2} + (k_3)_{i,k} \eta_{i,k+1/2} \frac{u^{n+\sigma/2}_{i,k+1} - u^{n+\sigma/2}_{i,k}}{h_z^2} -$$

$$-(k_4)_{i,k} \eta_{i,k-1/2} \frac{u^{n+\sigma/2}_{i,k} - u^{n+\sigma/2}_{i,k-1}}{h_z^2} + \left((k_3)_{i,k} - (k_4)_{i,k}\right) \frac{\tau_x}{\rho h_z}. \tag{5.2.9}$$

For boundary nodes located at the bottom of the region, equation (5.2.7):

$$(k_0)_{i,k} \frac{w^{n+\sigma}_{i,k} - w^n_{i,k}}{\tau} + (k_1)_{i,k} u^n_{i+1/2,k} \frac{w^{n+\sigma/2}_{i+1,k} - w^{n+\sigma/2}_{i,k}}{2h_x} +$$

$$+(k_2)_{i,k} u^n_{i-1/2,k} \frac{w^{n+\sigma/2}_{i,k} - w^{n+\sigma/2}_{i-1,k}}{2h_x} + (k_3)_{i,k} w^n_{i,k+1/2} \frac{w^{n+\sigma/2}_{i,k+1} - w^{n+\sigma/2}_{i,k}}{2h_z} +$$

$$+(k_4)_{i,k} w^n_{i,k-1/2} \frac{w^{n+\sigma/2}_{i,k} - w^{n+\sigma/2}_{i,k-1}}{2h_z} = (k_1)_{i,k} \mu_{i+1/2,k} \frac{w^{n+\sigma/2}_{i+1,k} - w^{n+\sigma/2}_{i,k}}{h_x^2} -$$

$$-(k_2)_{i,k} \mu_{i-1/2,k} \frac{w^{n+\sigma/2}_{i,k} - w^{n+\sigma/2}_{i-1,k}}{h_x^2} + \left((k_1)_{i,k} - (k_2)_{i,k}\right) \frac{\tau_z}{\rho h_x} +$$

$$+(k_3)_{i,k} \eta_{i,k+1/2} \frac{w^{n+\sigma/2}_{i,k+1} - w^{n+\sigma/2}_{i,k}}{h_z^2} - (k_4)_{i,k} \eta_{i,k-1/2}$$

$$\frac{w^{n+\sigma/2}_{i,k} - w^{n+\sigma/2}_{i,k-1}}{h_z^2} + (k_0)_{i,k} g. \tag{5.2.10}$$

Let us turn to the second subtask of wave hydrodynamics. We construct a discrete finite-volume model on the basis of equation (5.1.15), taking into account the following boundary conditions:

$$\frac{\partial P(x,z,t)}{\partial n} = \alpha_{p,n} P + \beta_{p,n}, \quad u(x,z,t) = u^{(0)}, \quad w(x,z,t) = w^{(0)}. \tag{5.2.11}$$

Integrate the equation (5.1.15) over the domain D_0:

$$\iint_{D_0}\frac{\partial^2 P}{\partial x^2}dxdz+\iint_{D_0}\frac{\partial^2 P}{\partial z^2}dxdz=\frac{\rho}{\tau}\left(\iint_{D_0}\frac{\partial u^{n+\sigma}}{\partial x}dxdz+\iint_{D_0}\frac{\partial w^{n+\sigma}}{\partial z}dxdz\right).$$

(5.2.1 2)

As a result of final integration, we can obtain the following expression:

$$(k_1)_{i,k}\frac{P_{i+1,k}-P_{i,k}}{h_x^2}-(k_2)_{i,k}\frac{P_{i,k}-P_{i-1,k}}{h_x^2}-\frac{\alpha_{p,x}P_{i,k}+\beta_{p,x}}{h_x}\left((k_1)_{i,k}-(k_2)_{i,k}\right)+$$

$$+(k_3)_{i,k}\frac{P_{i,k+1}-P_{i,k}}{h_z^2}-(k_4)_{i,k}\frac{P_{i,k}-P_{i,k-1}}{h_z^2}-\frac{\alpha_{p,z}P_{i,k}+\beta_{p,z}}{h_z}\left((k_3)_{i,k}-(k_4)_{i,k}\right)=$$

$$=\frac{\rho}{\tau}\frac{(k_1)_{i,k}u_{i+1/2,k}^{n+\sigma}-(k_2)_{i,k}u_{i-1/2,k}^{n+\sigma}-\left((k_1)_{i,k}-(k_2)_{i,k}\right)(u^{(0)})_{i,k}}{h_x}+$$

$$+\frac{\rho}{\tau}\frac{(k_3)_{i,k}w_{i,k+1/2}^{n+\sigma}-(k_4)_{i,k}w_{i,k-1/2}^{n+\sigma}-\left((k_3)_{i,k}-(k_4)_{i,k}\right)(w^{(0)})_{i,k}}{h_z}. \quad (5.2.13)$$

Next, we integrate the boundary condition on the free surface of the liquid (5.1.19) for the vertical component of the particle velocity over the region $D:\left\{\,(x,z)\in D_0;\;\; t\in\left(t^n,t^{n+1}\right)\right\}$, as a result we obtain:

$$(w^{(0)})_{i,k}=\frac{P_{i,k}^{n+1}-P_{i,k}^{n}}{g\rho\tau}. \quad (5.2.14)$$

In expression (5.2.13), the pressure field will be considered implicitly, in the case of boundary nodes at the bottom and surface of the region, as well as internal nodes, this expression will take the following form:

$$(k_1)_{i,k}\frac{P_{i+1,k}^{n+1}-P_{i,k}^{n+1}}{h_x^2}-(k_2)_{i,k}\frac{P_{i,k}^{n+1}-P_{i-1,k}^{n+1}}{h_x^2}+$$

$$+(k_3)_{i,k}\frac{P_{i,k+1}^{n+1}-P_{i,k}^{n+1}}{h_z^2}-(k_4)_{i,k}\frac{P_{i,k}^{n+1}-P_{i,k-1}^{n+1}}{h_z^2}=$$

$$=\frac{\rho}{\tau}\left(\frac{(k_1)_{i,k}u^{n+\sigma}_{i+1/2,k}-(k_2)_{i,k}u^{n+\sigma}_{i-1/2,k}}{h_x}+\frac{(k_3)_{i,k}w^{n+\sigma}_{i,k+1/2}-(k_4)_{i,k}w^{n+\sigma}_{i,k-1/2}}{h_z}\right)-$$

$$+\frac{\left((k_3)_{i,k}-(k_4)_{i,k}\right)\left(P^{n+1}_{i,k}-P^{n}_{i,k}\right)}{gh_z\tau^2}H\left((k_3)_{i,k}-(k_4)_{i,k}\right), \tag{5.2.15}$$

where $H(x)$- Heaviside function. We shall consider nodes for which the surface nodes are $k_3 > k_4$.

The expression (5.2.13) for the boundary nodes at the lateral boundary, taking into account the rise of the liquid level, takes the form:

$$(k_1)_{i,k}\frac{P_{i+1,k}-P_{i,k}}{h_x^2}-(k_2)_{i,k}\frac{P_{i,k}-P_{i-1,k}}{h_x^2}+(k_3)_{i,k}\frac{P_{i,k+1}-P_{i,k}}{h_z^2}-$$

$$-(k_4)_{i,k}\frac{P_{i,k}-P_{i,k-1}}{h_z^2}-\frac{\alpha_{p,x}}{h_x}\left((k_1)_{i,k}-(k_2)_{i,k}\right)=$$

$$=\frac{\rho}{\tau}\left(\frac{(k_1)_{i,k}u^{n+\sigma}_{i+1/2,k}-(k_2)_{i,k}u^{n+\sigma}_{i-1/2,k}}{h_x}+\frac{(k_3)_{i,k}w^{n+\sigma}_{i,k+1/2}-(k_4)_{i,k}w^{n+\sigma}_{i,k-1/2}}{h_z}\right)-$$

$$-\frac{\left((k_3)_{i,k}-(k_4)_{i,k}\right)\left(P^{n+1}_{i,k}-P^{n}_{i,k}\right)}{gh_z\tau^2}H\left((k_3)_{i,k}-(k_4)_{i,k}\right). \tag{5.2.16}$$

Next, proceed to the third subtask of wave hydrodynamics. We integrate the equation (5.1.13) over the domain D_0:

$$\iint_{D_0}\frac{u^{n+1}-u^{n+\sigma}}{\tau}dxdz=-\frac{1}{\rho}\iint_{D_0}\frac{\partial P}{\partial x}dxdz\,. \tag{5.2.17}$$

As a result of integration, the following expression is obtained (the pressure field is on the time layer $n + 1$):

$$(k_0)_{i,k}\frac{u^{n+1}_{i,k}-u^{n+\sigma}_{i,k}}{\tau}=-\left((k_1)_{i,k}\frac{P^{n+1}_{i+1,k}-P^{n+1}_{i,k}}{2h_x\rho}+(k_2)_{i,k}\frac{P^{n+1}_{i,k}-P^{n+1}_{i-1,k}}{2h_x\rho}\right). \tag{5.2.18}$$

Similarly, one can obtain a discrete analog of equation (5.1.11):

$$(k_0)_{i,k}\frac{w_{i,k}^{n+1}-w_{i,k}^{n+\sigma}}{\tau}=-\left((k_3)_{i,k}\frac{P_{i,k+1}^{n+1}-P_{i,k}^{n+1}}{2h_z\rho}+(k_4)_{i,k}\frac{P_{i,k}^{n+1}-P_{i,k-1}^{n+1}}{2h_z\rho}\right). \tag{5.2.19}$$

We introduce the «masks» of the boundary conditions: m_1, m_2. Parameter $m_1 = 1$, if the node (i,k) belongs to the set of boundary nodes located in the bottom area, otherwise $m_1 = 0$. Parameter $m_2 = 1$, if the node (i,k) belongs to a lateral boundary, otherwise it takes a value $m_2 = 1$.

Taking into account the introduced designations, the discrete model will be represented by the following equations:

- grid equation for calculating the component of the velocity vector $u_{i,k}$:

$$(k_0)_{i,k}\frac{u_{i,k}^{n+\sigma}-u_{i,k}^{n}}{\tau}+(k_1)_{i,k}u_{i+1/2,k}^{n}\frac{u_{i+1,k}^{n+\sigma/2}-u_{i,k}^{n+\sigma/2}}{2h_x}+(k_2)_{i,k}u_{i-1/2,k}^{n}\frac{u_{i,k}^{n+\sigma/2}-u_{i-1,k}^{n+\sigma/2}}{2h_x}+$$

$$+(k_3)_{i,k}w_{i,k+1/2}^{n}\frac{u_{i,k+1}^{n+\sigma/2}-u_{i,k}^{n+\sigma/2}}{2h_z}+(k_4)_{i,k}w_{i,k-1/2}^{n}\frac{u_{i,k}^{n+\sigma/2}-u_{i,k-1}^{n+\sigma/2}}{2h_z}=$$

$$=(k_1)_{i,k}\mu_{i+1/2,k}\frac{u_{i+1,k}^{n+\sigma/2}-u_{i,k}^{n+\sigma/2}}{h_x^2}-(k_2)_{i,k}\mu_{i-1/2,k}\frac{u_{i,k}^{n+\sigma/2}-u_{i-1,k}^{n+\sigma/2}}{h_x^2}+$$

$$+(k_3)_{i,k}\eta_{i,k+1/2}\frac{u_{i,k+1}^{n+\sigma/2}-u_{i,k}^{n+\sigma/2}}{h_z^2}-(k_4)_{i,k}\eta_{i,k-1/2}\frac{u_{i,k}^{n+\sigma/2}-u_{i,k-1}^{n+\sigma/2}}{h_z^2}+$$

$$+\left((k_3)_{i,k}-(k_4)_{i,k}\right)\frac{\tau_x}{\rho h_z}(m_1)_{i,k}, \tag{5.2.20}$$

- grid equation for calculating the component of the velocity vector $w_{i,k}$:

$$(k_0)_{i,k}\frac{w_{i,k}^{n+\sigma}-w_{i,k}^{n}}{\tau}+(k_1)_{i,k}u_{i+1/2,k}^{n}\frac{w_{i+1,k}^{n+\sigma/2}-w_{i,k}^{n+\sigma/2}}{2h_x}+$$

$$+(k_2)_{i,k} u^n_{i-1/2,k} \frac{w^{n+\sigma/2}_{i,k} - w^{n+\sigma/2}_{i-1,k}}{2h_x} + (k_3)_{i,k} w^n_{i,k+1/2} \frac{w^{n+\sigma/2}_{i,k+1} - w^{n+\sigma/2}_{i,k}}{2h_z} +$$

$$+(k_4)_{i,k} w^n_{i,k-1/2} \frac{w^{n+\sigma/2}_{i,k} - w^{n+\sigma/2}_{i,k-1}}{2h_z} = (k_1)_{i,k} \mu_{i+1/2,k} \frac{w^{n+\sigma/2}_{i+1,k} - w^{n+\sigma/2}_{i,k}}{h_x^2} -$$

$$-(k_2)_{i,k} \mu_{i-1/2,k} \frac{w^{n+\sigma/2}_{i,k} - w^{n+\sigma/2}_{i-1,k}}{h_x^2} + \left((k_1)_{i,k} - (k_2)_{i,k}\right) \frac{\tau_z}{\rho h_x} (m_1)_{i,k} +$$

$$+(k_3)_{i,k} \eta_{i,k+1/2} \frac{w^{n+\sigma/2}_{i,k+1} - w^{n+\sigma/2}_{i,k}}{h_z^2} - (k_4)_{i,k} \eta_{i,k-1/2} \frac{w^{n+\sigma/2}_{i,k} - w^{n+\sigma/2}_{i,k-1}}{h_z^2}$$

$$+(k_0)_{i,k} g. \tag{5.2.21}$$

- grid equations for calculating the pressure field:

$$(k_1)_{i,k} \frac{P_{i+1,k} - P_{i,k}}{h_x^2} - (k_2)_{i,k} \frac{P_{i,k} - P_{i-1,k}}{h_x^2} + (k_3)_{i,k} \frac{P_{i,k+1} - P_{i,k}}{h_z^2} -$$

$$-(k_4)_{i,k} \frac{P_{i,k} - P_{i,k-1}}{h_z^2} - \frac{\alpha (m_2)_{i,k}}{h_x} \left((k_1)_{i,k} - (k_2)_{i,k}\right) =$$

$$= \frac{\rho}{\tau} \left(\frac{(k_1)_{i,k} u^{n+\sigma}_{i+1/2,k} - (k_2)_{i,k} u^{n+\sigma}_{i-1/2,k}}{h_x} + \frac{(k_3)_{i,k} w^{n+\sigma}_{i,k+1/2} - (k_4)_{i,k} w^{n+\sigma}_{i,k-1/2}}{h_z} \right) -$$

$$- \frac{\left((k_3)_{i,k} - (k_4)_{i,k}\right)\left(P^{n+1}_{i,k} - P^n_{i,k}\right)}{g h_z \tau^2} H\left((k_3)_{i,k} - (k_4)_{i,k}\right). \tag{5.2.22}$$

- grid equations for the refinement of the velocity field with respect to pressure:

$$(k_0)_{i,k} \frac{u^{n+1}_{i,k} - u^{n+\sigma}_{i,k}}{\tau} = -\left((k_1)_{i,k} \frac{P^{n+1}_{i+1,k} - P^{n+1}_{i,k}}{2h_x \rho} + (k_2)_{i,k} \frac{P^{n+1}_{i,k} - P^{n+1}_{i-1,k}}{2h_x \rho} \right), \tag{5.2.23}$$

$$(k_0)_{i,k}\frac{w_{i,k}^{n+1}-w_{i,k}^{n+\sigma}}{\tau}=-\left(\begin{array}{l}(k_3)_{i,k}\dfrac{P_{i,k+1}^{n+1}-P_{i,k}^{n+1}}{2h_z\rho}\\+(k_4)_{i,k}\dfrac{P_{i,k}^{n+1}-P_{i,k-1}^{n+1}}{2h_z\rho}\end{array}\right). \qquad (5.2.24)$$

Thus, it can be noted that a finite-volume model of the problem is constructed, which is represented by the equations (5.2.20) and (5.2.21), (5.2.22), (5.2.23) and (5.2.24).

5.3 Discrete Model Research

Conservative conditions require the preservation of certain integral relations, which are valid for the initial differential equations. Consequently, the difference scheme is conservative if the integral conservation law is observed for its grid analog.

The conservativeness of the discrete model is investigated. The method of splitting the initial equations into physical processes is an additive scheme; it is stable and guarantees the fulfillment of the mass balance (continuity equation). It is established that the integral law of conservation of momentum is observed for the grid analog of the difference equation.

The approximation error of the finite-difference scheme was estimated to be $O(\tau+h_x^2+h_z^2)$. A stability analysis was performed for the problem using the maximum principle and limitations on time and spatial steps were obtained.

Method for Solving the Problem

Numerical solution of two-dimensional and especially three-dimensional problems of mathematical physics often requires the solution of large and super-large SLAE with the number of unknowns more than 1000. Direct methods for solving SLAE, such as the Gaussian method and its modification, usually have a cubic dependence of the number of arithmetic operations on the size of the SLAE matrix. At N>1000, this dependence leads to an unrealistically long solution time. In addition, the decision time is disproportionately increased due to the insufficient amount of RAM for storing task data.

Iterative methods for solving SLAE are much more economical, both in terms of the computer time of the solution and the use of RAM. Thus, if the iteration method is rapidly converging with the number of iterations

$m << N$, then the solution time, which is proportional to the square of the size of the matrix $\sim mN^2$, turns out to be substantially smaller. In addition, it is required to store in RAM, as a rule, only one matrix.

The most algorithmically simple among the iterative methods are stationary iterative methods, such as the optimal simple iteration method and the relaxation method. To solve the discrete equations of the problem of the run-up of surface wave (5.2.20) and (5.2.21), (5.2.22), (5.2.23) and (5.2.24) we will use the over relaxation method [Samarsky *et al.*, 1989]. The merits of the iterative method of upper relaxation can also be attributed to the fact that, when implemented on a computer, the computation algorithm has a simple form and allows us to use only one array for an unknown vector.

The Algorithm of the Program "2DbayWaves"

To calculate the two-dimensional velocity field of motion and the pressure field of an aqueous medium, the software code "2DBayWaves" was created on the run-up of nonlinear surface gravity wave on the shore of shallow water areas [Abbasov, Semenov, Tsarevsky, 2012].

The software code 2DBayWaves was compiled and implemented in C ++ in the Windows Vista operating system. The developed program takes into account the turbulent exchange, complex geometry of the bottom and shoreline, friction against the bottom, and ensures the following operations:

- calculation of the velocity field without taking pressure into account;
- calculation of the pressure field;
- calculation of a two-dimensional velocity field.

The structure of the 2DBayWaves software code consists of the following blocks:

- input block for calculating speed and pressure, defining the bottom relief geometry, mask of the boundary conditions and initial values of the cells' fullness;
- block for constructing grid equations for the velocity field without allowance for pressure;
- block for constructing grid equations for the pressure field;
- block for calculating the fullness of cells;
- block for calculating the field of a two-dimensional velocity vector (on the next time layer);
- unit for calculating the grid equations by the over relaxation method;
- block for output of pressure and fullness of cells.

1. Input specification:
   - the size of the grid $\varpi_h = \{N_t\tau = T,\ N_x h_x = l_x,\ N_z h_z = l_z\}$;
   - N_t – upper time limit, step by time τ;
   - N_x, N_z – borders on space, steps in space h_x, h_z;
   - setting the initial pressure distribution on the left side boundary;
   - setting the depth of the reservoir H;
   - setting the geometry of the bottom line and shoreline.
2. Calculation of the velocity components of the particles of the medium without taking into account the pressure:
   - after numbering the elements and calculating the numbers of neighboring elements in all coordinate directions, the coefficients of the grid equations for the nodes in the vicinity of the template center are calculated, the program uses one-dimensional arrays (element number i,k);
   - the calculation of the coefficients of the grid equations and the right sides is performed in accordance with the representation of the grid equation in the canonical form;
   - solution of the grid equations by an iterative method of over relaxation.
3. Calculation of the pressure field,
   - for all nodes, the coefficients and the right side of the grid equation for calculating the pressure are written;
   - calculation of the pressure field on the basis of calculation of SLAE by the method of over relaxation.
4. Calculation of the fullness of cells.
5. Refinement of the velocity field by pressure.
6. Output of pressure field data.
7. Then increase the counter by time by one.
8. If the time is shorter than the exit time, return to step 2.
9. The end.

5.4 Results of Two-Dimensional Numerical Modeling and Their Analysis

After considering the calculation algorithm, we turn to the results of numerical simulation of the run-up of nonlinear surface gravity waves onto shore formations of shallow water areas.

It is noteworthy that the shallow-water condition for the Navier–Stokes equation need not be satisfied. Therefore, the initial parameters of surface gravity waves will have a shallow-water parameter within $kH \geq 1$, but with further approach to the shore, the depth will be reduced and the surface wave will meet the shallow-water conditions.

According to the statement of the problem, a layer of viscous incompressible fluid approaches the shore, while the bottom line passes to the shoreline. In accordance with the geometry of the problem, the axis Ox of the coordinate system coincides with the surface of the unperturbed liquid and is directed toward the shore, the axis Oz is directed vertically upwards. At the initial instant of time, the liquid is at rest. At a certain distance from the coast at the point x = 0, a perturbation in the form of a pressure pulse, varying according to the harmonic law, is defined on the left lateral boundary. It is necessary to follow the subsequent movement of the water.

Before the 2DBayWaves software code is started, the sizes of the rectangular grid (N_x, N_z–the border along the space) and the number of iterations over time (N_t–the upper time limit), are set, as well as the parameters of the surface gravity wave (amplitude – a, wavelength – λ).

Further, a profile of the bottom and the shoreline on the basis of a spline on four points is constructed. In this case, the depth of the reservoir is set to the liquid level at rest. After setting this data, the window for visualizing the pressure change of the surface gravity wave is launched, which is shown in Figure 5.4.1.

As a shallow-water model, we use the hydrophysical conditions of the Taganrog Bay of the Azov Sea. In our case, the length of the shallow-water area is determined by the grid size. Also, to ensure that the physical process is correctly described, the nonlinear effects in the propagation of surface waves must be accumulation gradually. The grid size will be 200x600, and the water surface level is 100 vertically. The bay should be as extended along the Ox-axis as to involve at least two wavelengths.

In view of the bay depth of $H \leq 5$m, the segment of the water area will have the following geometric dimensions: a length of 30m, a vertical length of 10m (from the bottom), a horizontal step of $h_x = 0.05$m, and a vertical step of $h_z = 0.05$m. The length of the surface wave will be $\lambda \leq 15$m, the initial values of the shallow-water parameter will be in the range $1 \leq kH \leq 10$, the rate of surface wave propagation will depend on frequency based on a dispersion relation.

We simulate the coastal slopes of different steepness in the conditions of the bay. The bottom lines are shown in Figure 5.4.2. They are based on graphs of power functions. The depth decreases from $H = 5$m to zero, the slope steepness does not exceed 0.2^0. The bottom line $s2(x)$ is more flat, and the line $s1(x)$ is steeper along the shore. Also, we will use a uniformly (linearly) rising bottom line.

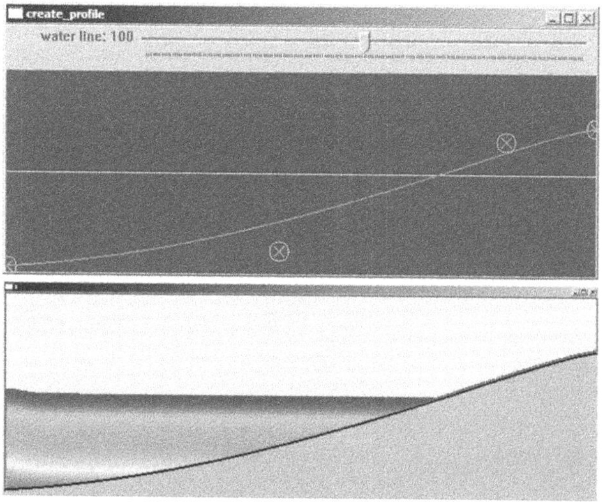

Figure 5.4.1 Windows for constructing a bottom profile and visualizing the pressure field.

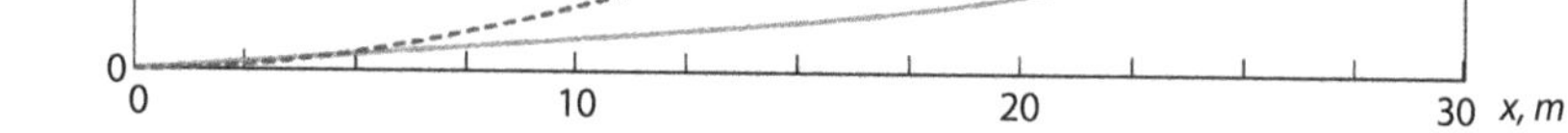

Figure 5.4.2 Lines of the bottom according to analytic expressions: $s1(x) = ax + bx^8$; $s2(x) = ax + bx^2 - cx^3$;

Figure 5.4.3÷5.4.6 shows the dynamics of arrival of nonlinear surface gravity waves on a flat coastal slope $s2(x)$. On the left lateral boundary, harmonic disturbance as a pressure pulse is specified. The surface gravity wave satisfies the initial shallow-water requirements $H/\lambda < 1/2$, and its length is twice the depth [Whitham, 1974], [*Dynamics of Channel Flows*, 1994].

As approaching to the coast, the bay depth is reduced, the wave begins to touch the bottom and, thus, the impact of nonlinear effects increases. This leads to a steepened front of the surface wave crest. The trajectories of medium particles on the wave forefront become vertical, the forefront becomes sheer, and the wave is broken. This process can be observed at time step of t = 3.6s (Figure 5.4.3). When breaking, the wave falls apart,

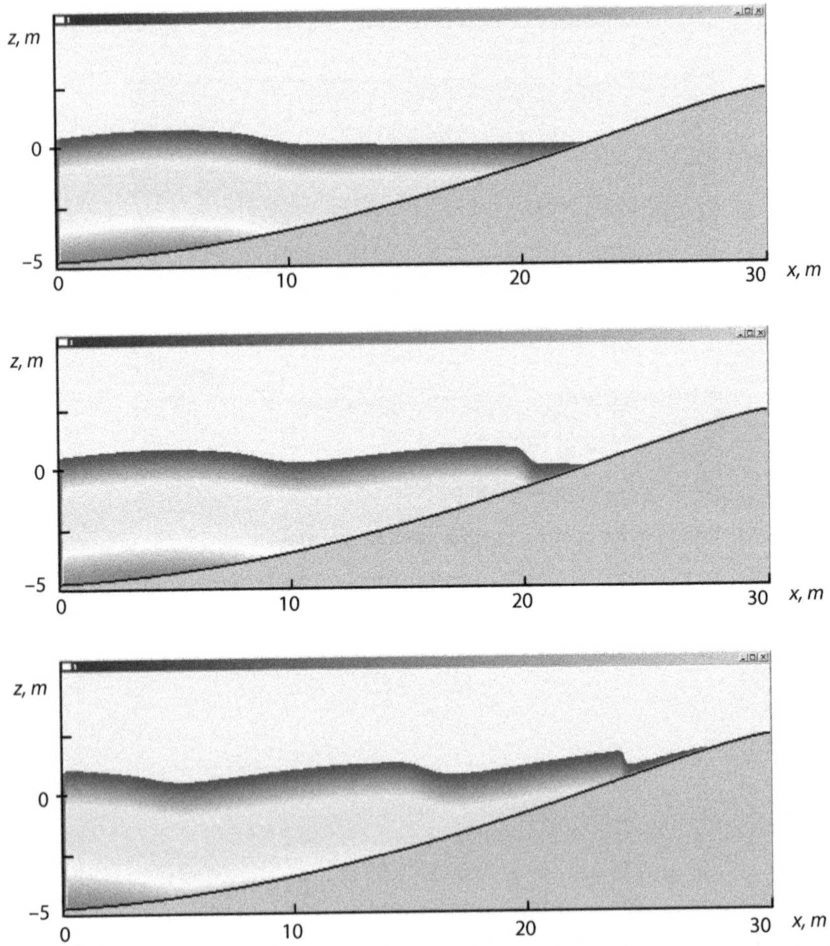

Figure 5.4.3 Dynamics of changes in the profile of incident surface gravity wave on the coastal slope (gentle rise), the initial parameters of the wave: $f = 0.39$Hz; $\lambda = 10$m; $c = 4$m/s; H = 5m; $a = 0.5$m; $kH = 3.14$; $\varepsilon = 0.1$; time (from top to bottom) $t = 1.6$s; $t = 3.6$s; $t = 6.8$s.

which leads to a reduced wave height and an underflooded coastal slope. Later, during the wave run-up, the water flow begins to run back along the inclined slope. The wave that runs back knocks the next wave, which further enhances the steepening and accelerates the breaking of the next wave.

Simultaneously, the rollback wave raises the initial liquid level relative to the resting state, enhancing the penetration of the run-up wave onto the land. This process is clearly demonstrated in Figure 5.4.4, especially the lower visualization. For this visualization, the wave height is the largest

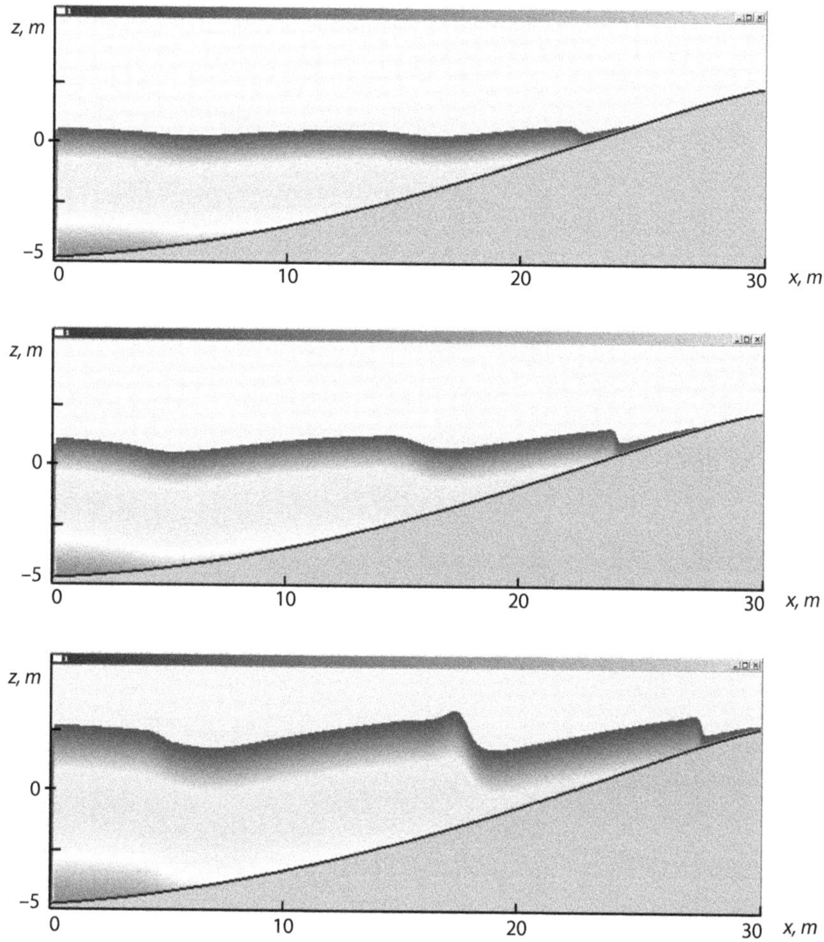

Figure 5.4.4 A run-up on the coastal slope of surface gravity waves of different amplitude, the initial wave parameters: $f = 0.39$Hz; $\lambda = 10$m; $c = 4$m/s; H = 5m; $kH = 3.14$; time (from top to bottom) 1) $t = 7$s; $a = 0.25$m; $\varepsilon = 0.05$; 2) $t = 6.8$s; $a = 0.5$m; $\varepsilon = 0.1$; 3) $t = 6.9$s; $a =$ 1m; $\varepsilon = 0.2$.

(a = 1m), and the dynamics of distortions of the wave profile is the most pronounced.

Figure 5.4.5 shows the process of run-up on the same coastal slope of surface waves of different lengths. Here, each wave has its initial shallow water parameter in the interval $2.1<kH<12.6$. For the shortest wave $\lambda = 2.5$m the deep-water conditions $kH = 12.6$ are actually observed. Consequently, for this wave, the nonlinear distortions are least appreciable because of the influence of depth. For the longest wave $\lambda = 15$m despite the smallest of all

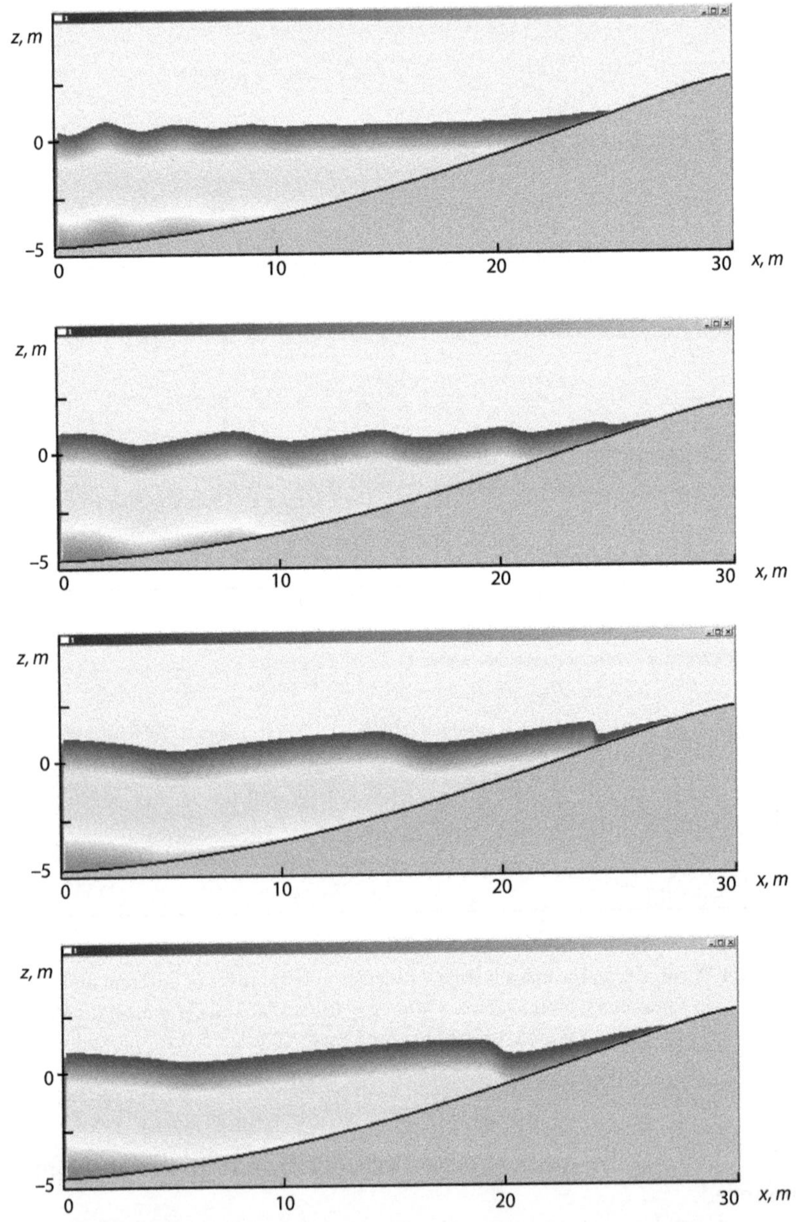

Figure 5.4.5 The run-up on the coastal slope of surface gravity waves of different lengths, the initial wave parameters: H = 5m; a = 0.5m; ε = 0.1; (from top to bottom),

1)	f = 0.79Hz;	λ = 2.5m;	c = 2m/s;	kH = 12.6;	t = 5.2s;
2)	f = 0.56Hz;	λ = 5m;	c = 2.8m/s;	kH = 6.3;	t = 5.6s;
3)	f = 0.39Hz;	λ = 10m;	c = 4m/s;	kH = 3.14;	t = 6.8s;
4)	f = 0.32Hz;	λ = 15m;	c = 4.8m/s;	kH = 2.1;	t = 6.9s.

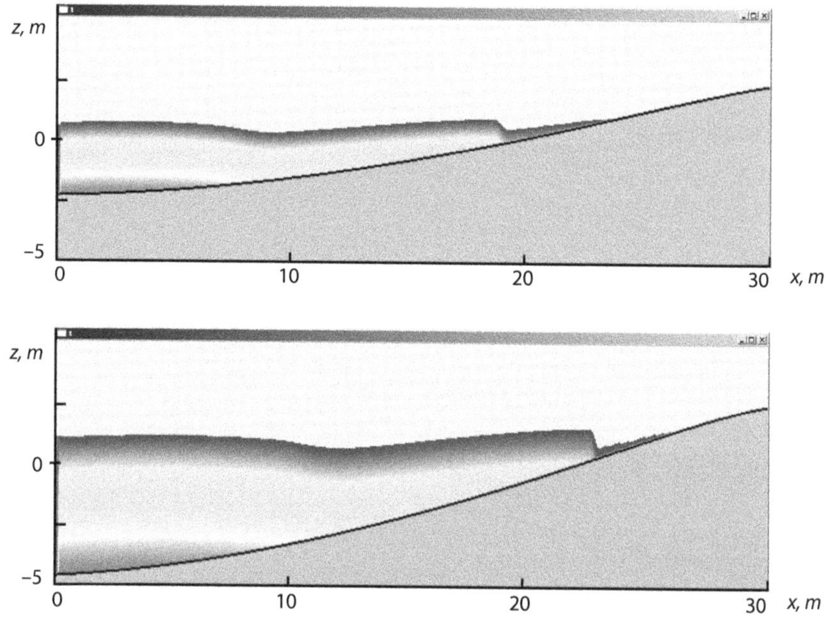

Figure 5.4.6 A run-up on the coastal slopes of different depths of the surface gravity wave, the initial wave parameters, from top to bottom: 1) $f = 0.28$Hz; $\lambda = 15$m; $c = 4.3$m/s; H = 2.5m; $a = 0.5$m; $kH = 1$; $\varepsilon = 0.4$; $t = 8$s; 2) $f = 0.32$Hz; $\lambda = 15$m; $c = 4.8$m/s; H = 5m; $a = 0.5$m; $kH = 2.1$; $\varepsilon = 0.1$; $t = 8$s.

waves, the initial slope is nonlinear distortion are most pronounced, since it is the shallowest $kH = 2.1$.

It can also be noted that for the shortest wave $\lambda = 2.5$m the process of run-up has its own peculiarities. Here, nonlinear distortions of the wave profile do not appear, the coastline runs in the form of a continuous stream of water. With increasing length of the surface wave, distortion of the profile due to nonlinearity amplifies, and the crest of the wave is broken.

In Figure 5.4.6. the run-up of the surface gravity wave on the coastal slopes with different initial depths is presented. Reducing the depth of the bay for the same wave leads to an increase in nonlinearity, so that the wave breaking faster (upper visualization).

In Figure 5.4.7. the dynamics of the change in the profile of the surface gravity wave, which is incident on the coastal slope with a linear rise, is given.

In Figure 5.4.8. the dynamics of the change in the profile of the surface gravity wave, which runs into the coastal slope with a steep rise (the bottom line $s1(x)$) is given.

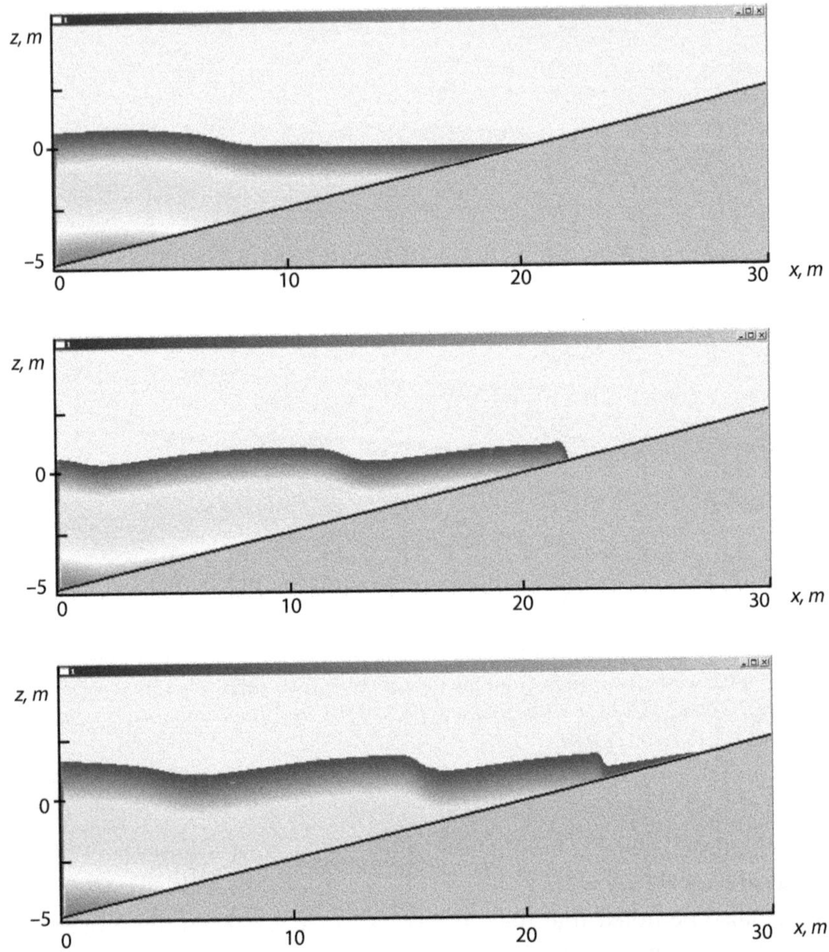

Figure 5.4.7 Dynamics of the change in the profile of the surface gravity wave incident on the coastal slope (linear rise), the initial parameters of the wave: $f = 0.39$Hz; $\lambda = 10$m; $c = 4$m/s; H = 5m; $a = 0.5$m; $kH = 3.14$; $\varepsilon = 0.1$; time (from top to bottom) $t = 1.4$s; $t = 4.3$s; $t = 8.9$s.

From the comparison, it can be noted that a linearly rising bottom enhances nonlinear distortions, rather than a slowly rising bottom with a steep rise at the shore. The surface gravity wave is broken on the shore without any initial distortion of the profile. However, because of the vertical coastal wall, a sufficiently tangible reflected run back wave is formed.

Because of the imposition of incident and reflected waves, their interference occurs. This process is shown in Figure 5.4.9, the position of the

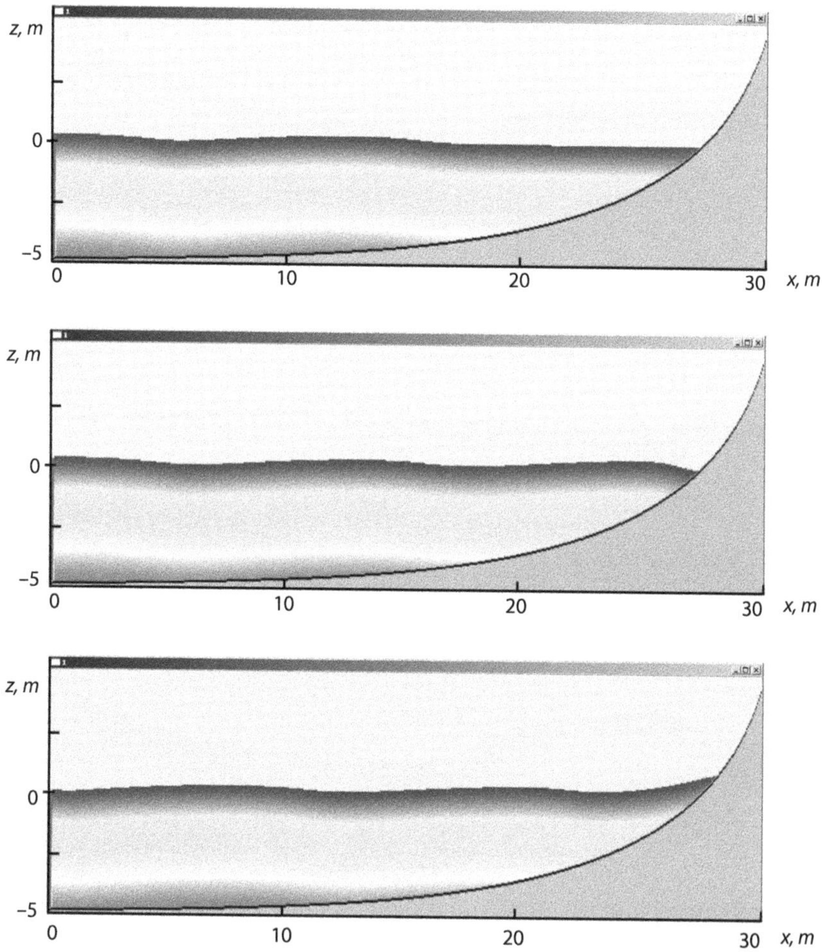

Figure 5.4.8 Dynamics of changes in the profile of a surface gravity wave that hits the coastal slope (steep rise), initial wave parameters: $f = 0.39$Hz; $\lambda = 10$m; $c = 4$m/s; H = 5m; $a = 0.5$m; $kH = 3.14$; $\varepsilon = 0.1$; time (from top to bottom) $t = 2.9$s; $t = 5.1$s; $t = 6.2$s.

reflected wave in the visualizations is indicated by an arrow. On the average visualization, due to superposition, the waves are mutually quenched, which results in the appearance of a continuous stream with one forefront.

Figure 5.4.10 shows the process of run-up of the same surface wave on the coastal slopes of different steepness. From the comparison, it can be noted that the greatest distortion undergoes a wave profile on the coastal slope with a uniform rise, i.e., the depth decreases faster than all the slopes (medium visualization).

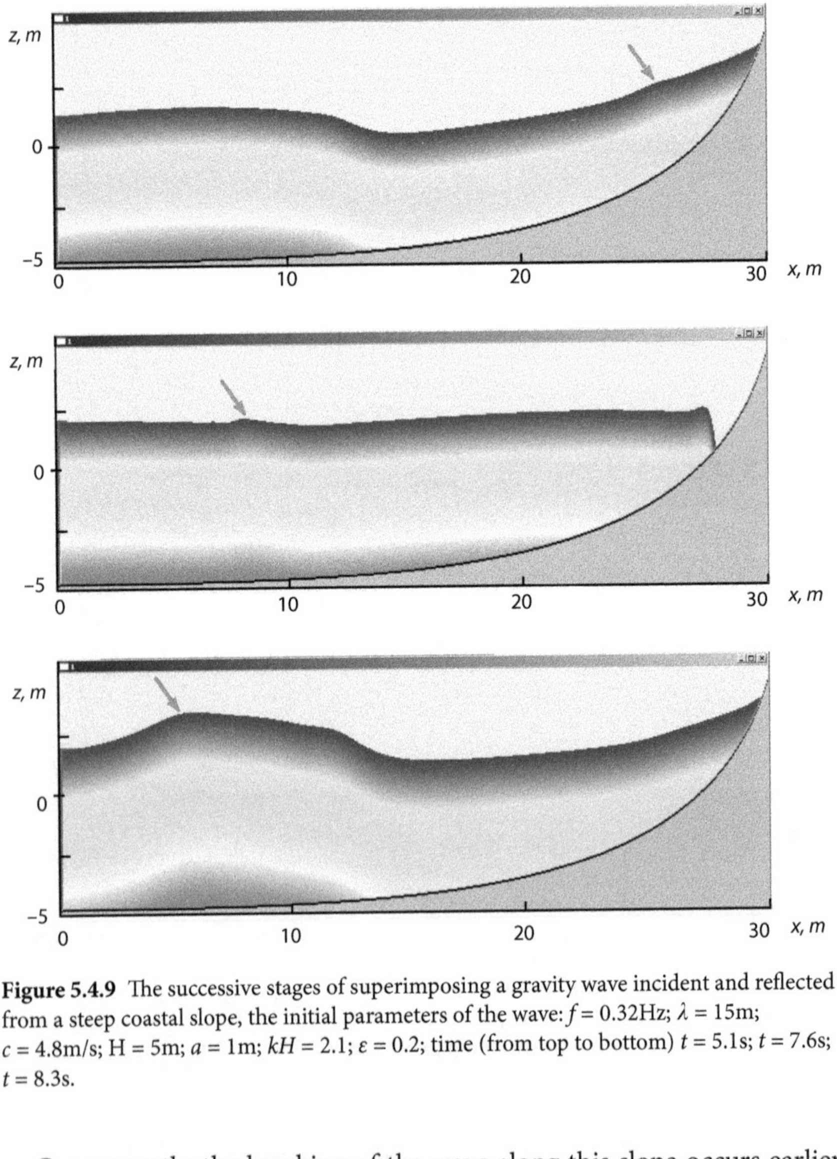

Figure 5.4.9 The successive stages of superimposing a gravity wave incident and reflected from a steep coastal slope, the initial parameters of the wave: $f = 0.32$Hz; $\lambda = 15$m; $c = 4.8$m/s; H = 5m; $a = 1$m; $kH = 2.1$; $\varepsilon = 0.2$; time (from top to bottom) $t = 5.1$s; $t = 7.6$s; $t = 8.3$s.

Consequently, the breaking of the wave along this slope occurs earlier than the rest. The smallest distortions of the wave profile are observed along the slope with a steep rise near the shore (upper).

On the offshore shelves in the World Ocean, water areas with an extensive shallow coastal area are often observed. During storm surges or in the event of a tsunami, a complete flooding of the coastal zone and land may occur. Figure 5.4.11 in the long-wave approximation modeled the run-up of the surface gravity wave to the water area of the shallow water.

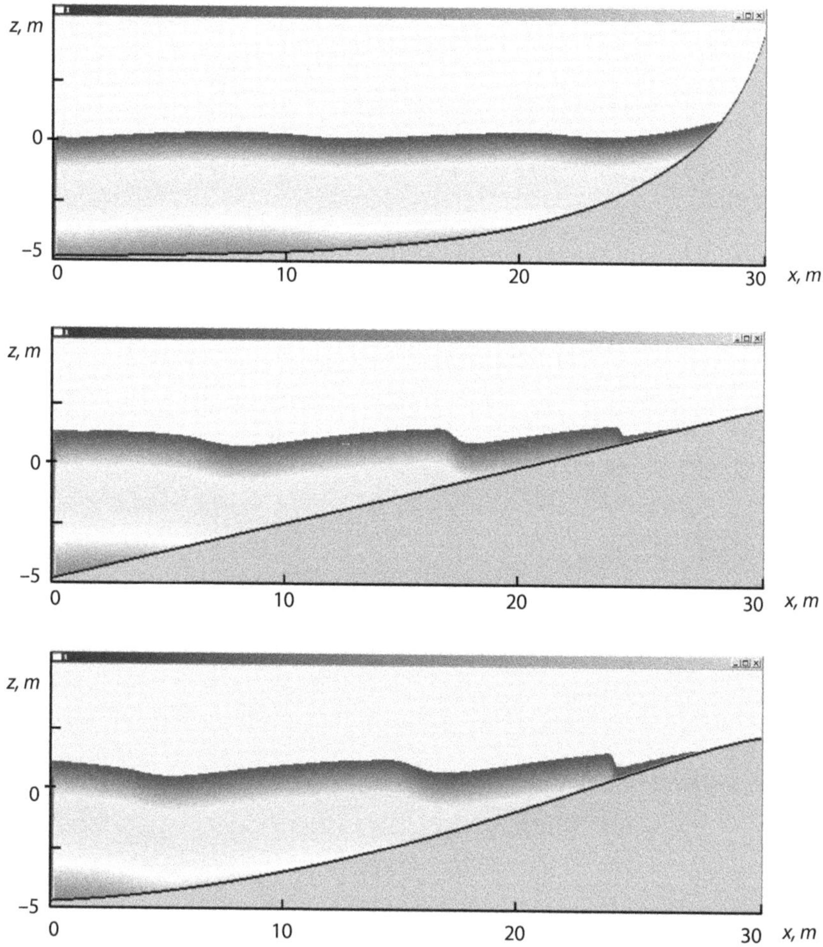

Figure 5.4.10 Dynamics of changes in the profile of a surface gravity wave that runs into the coastal slopes with different steepness, initial wave parameters: $f = 0.39$Hz; $\lambda = 10$m; $c = 4$m/s; H = 5m; $a = 0.5$m; $kH = 3.14$; $\varepsilon = 0.1$; (from top to bottom) 1) steep rise $t = 6.2$s; 2) linear rise $t = 7.3$s; 3) loping rise $t = 6.8$s.

From Figure 5.4.11 it can be seen that, under the continuous action of surface gravity waves, there is a complete flooding of an extensive shallow water area. Accordingly, within the framework of our model, it is also possible to predict the impact of storm surges and tsunamis on the shallow water areas of the World Ocean.

The stability of the developed numerical model was verified using various initial parameters:

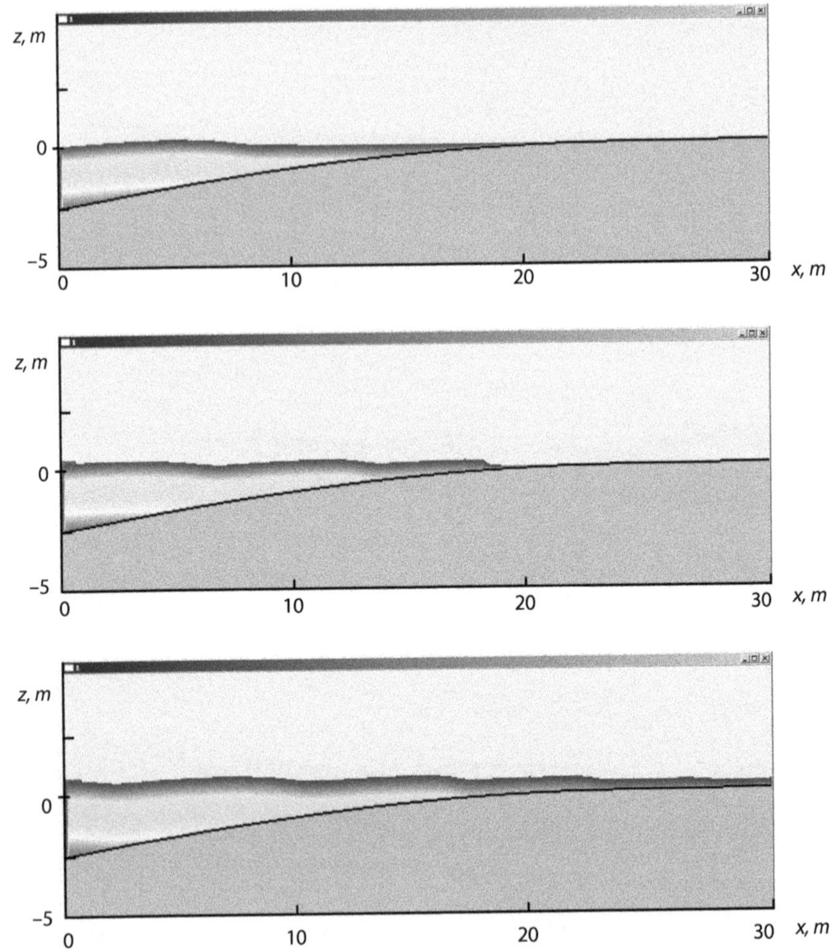

Figure 5.4.11 The dynamics of the change in the profile of the incident surface gravity wave and the complete flooding of the long shallow shore, the initial parameters of the wave: $f = 0.28$Hz; $\lambda = 15$m; $c = 4.3$m/s; H = 2.5m; $a = 0.25$m; $kH = 1$; $\varepsilon = 0.2$; time (from top to bottom) $t = 2.1$s; $t = 5.6$s; $t = 18.6$s.

- the extent of the water basin 30x10m; 30x12.7m; 15x5m; 45x15m;
- grid size: 300x100; 600x200; 650x270; 900x300 cells;
- grid step $h_x = h_z = 0.05$m; 0.047m; 0.03m; 0.1m;
- basin depth H = 2.5m; 5m; 7.5m;
- wavelength $\lambda = 2.5$m; 5m; 10m; 15m;
- wave amplitude $a = 0.25$m; 0.5m; 1m.

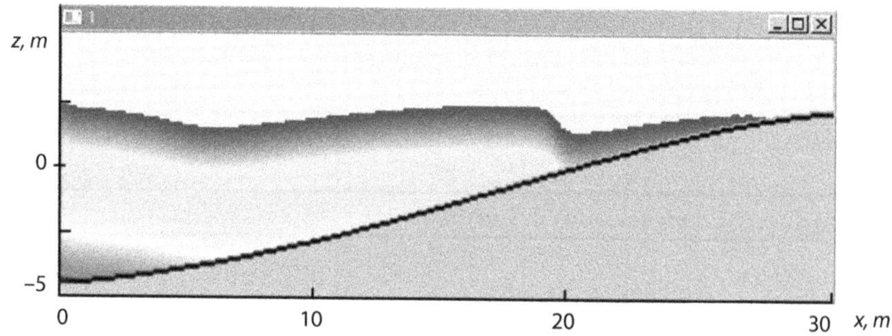

Figure 5.4.12 Run-up of the surface gravity wave on the coastal slope with initial parameters: λ = 10m; c = 4m/s; H = 5m; a = 0.5m; kH = 3.14; ε = 0.1; t = 6.5s (grid with a larger step size $h_x = h_z$ = 0.1m).

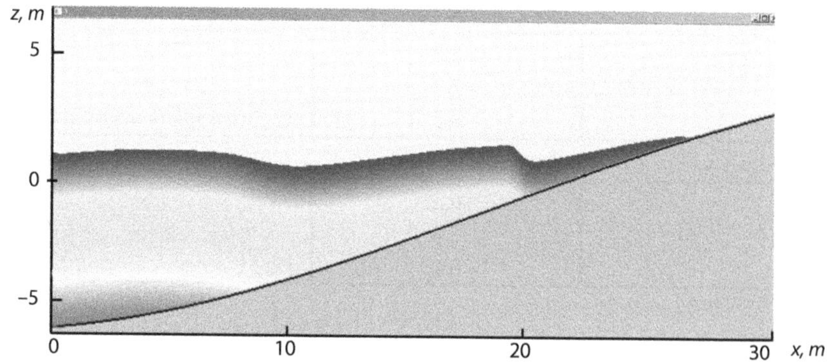

Figure 5.4.13 Run-up of surface gravity wave on the coastal slope with initial parameters: λ = 10m; c = 4m/s; H = 6.5m; a = 0.5m; kH = 4; ε = 0.07; t = 5.6s (grid step $h_x = h_z$ = 0.047m, size of area 30x12.7m).

Figures 5.4.12 and 5.4.13 show the process of run-up the surface wave for different values of the grid pitch and shallow water depth: grid spacing $h_x = h_z$ = 0.1m, size of the area 30x10m (Figure 5.4.12); grid step $h_x = h_z$ = 0.047m, the size of the area is 30x12.7m (Figure 5.4.13). It can be noted that for the new calculation parameters, the described wave processes are also observed, which indicates the preservation of the occurring physical processes.

To analyze the run-up of surface gravity waves quantitatively, the table gives some parameters used in the computations. They estimate the run-up height and distance as functions of the length and initial steepness ($2a/\lambda$) of the nonlinear surface wave. Such situations occur in shallow water during storm surges.

Table 5.1 Comparative parameters for the run-up of nonlinear surface gravity waves on a gentle slope for H = 5m, a hill height of 7.5m, basin area of 30 × 10m, and grid size 300 × 100.

steepness, $2a/\lambda$	run-up height H_{runup}, m	run-up distance L_{runup}, m
wavelength λ = 2.5 m; velocity c = 1.98 m/s		
0.2	6.1	3.1
0.4	6.5	5.5
0.8	7.3	9.2
wavelength λ = 5 m; velocity c = 2.78 m/s		
0.1	5.8	2.5
0.2	6.5	6.1
0.4	8.1	12.2 (*complete inundation*)
wavelength λ = 10 m; velocity c = 3.94 m/s		
0.05	5.8	3.5
0.1	7.2	7.4
0.2	8.5	12.2 (*complete inundation*)
wavelength λ = 15 m; velocity c = 4.77 m/s		
0.03	6.2	3.6
0.07	7.3	10.3
0.13	8.6	12.2 (*complete inundation*)

As the wavelength and amplitude (steepness) increase, the run-up distance increases as well and complete inundation is observed for some values. Therefore, this model makes it possible to predict the scale of inundation at the run-up of surface gravity waves.

Main Results

As a result of numerical simulation of the run-up of nonlinear surface gravity waves on shore formations of shallow water areas, it was established:

- with the approach to the shore, the depth of the bay decreases, the wave begins to sense the bottom, the influence of nonlinear effects increases. This leads to a steepening of the leading edge of the crest of the surface wave and in the future, it collapses. Further, the wave height decreases and the coastal slope (the run-up of wave) is flooded. A wave that run-back is knocking down the next wave, which further strengthens the steepening and accelerates the breaking of the next wave;

- an increase in the initial steepness of the surface wave leads to the wave breaking on the dry coastal zone, while, in the case of a less steep initial wave, the run-up occurs without wave breaking;
- for a shorter wavelength, the shallow water conditions are satisfied worse, the bottom caused nonlinear distortions of the wave profile decrease, and the wave run-up occurs in the form of a continuous water flow. As the length of the surface wave increases, the profile distortions due to the nonlinearity are enhanced and the wave crest breaks;
- A decrease in the depth of the bay for the same wave also leads to an increase in nonlinearity, as a result of which the breaking process accelerates;
- the bottom line profile influences on the breaking process when the run-up of surface wave. A linearly rising bottom enhances nonlinear distortions compared to a slowly rising bottom with a steep rise at the shore; the faster the depth decreases, the more the wave profile distorts and the more it breaks. With a slow decrease in depth, the surface wave can break on the shore without actually distorting the profile;
- In the case of a steep rise of the coastal slope, a reflected run-back wave is formed, which is superimposed on the incoming wave. In the case of overlap in the antiphase, there is a mutual quenching of the waves, which causes a continuous flow with a single forefront;
- the run-up of surface gravity waves and tsunami waves may lead to complete coastal inundation in a shallow water basin.

5.5 Discussion and Comparison of Results

To check the adequacy of the model developed, the results of two-dimensional numerical simulation of arriving nonlinear surface gravity waves on coastal structures were compared with existing numerical and experimental data.

Figure 5.5.1 presents the results of three models for numerical modeling of the run-up of nonlinear surface gravity waves a flat coastal slope.

Figure 5.5.1a shows the results of simulating the run-up and breaking of the surface wave on a flat slope on the basis of the Navier-Stokes equations of Japanese researchers [Kawasaki *et al.*, 2010]. When the wave breaks, the interaction between air and water, as well as the nonlinear effects

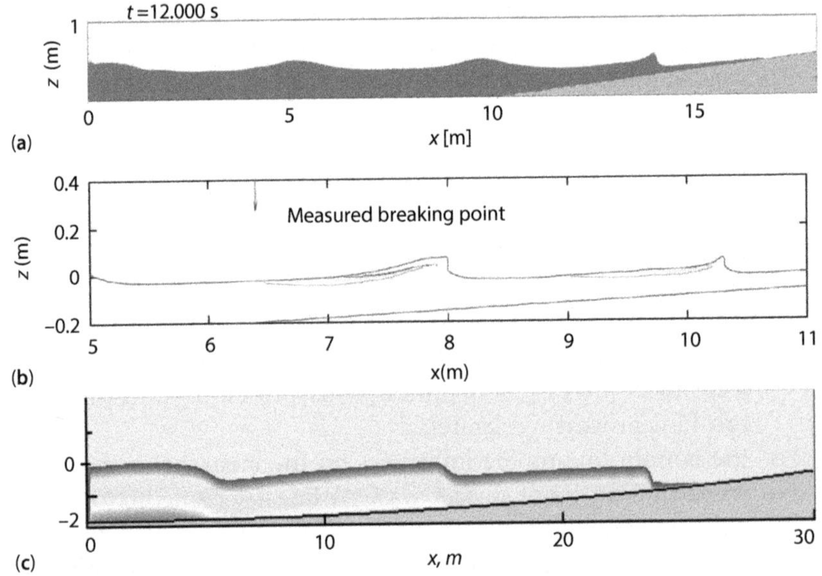

Figure 5.5.1 The comparison of run-up surface gravity wave on a flat coastal slope on the basis of different models: (a) λ = 5m; H = 0.5m; T = 2.0s; a = 0.15m; t = 12s [Kawasaki *et al.*, 2010]; (b) λ = 3m; H = 0.2m; T = 5.0s; a = 0.128m [Zhao *et al.*, 2004]; (c) λ = 10m; H = 2m; a = 1m; c = 3.6m/s; kH = 1.3; ε = 0.5; t = 5.4s (investigated model).

themselves, are taken into account. In the calculated area (grid 540x80), the wave excitation source is located at the point x = 0m, the slope of the coastline has a constant gradient of 1/15 in the section from x = 10m to x = 20m. The wave period is 2s, and the wave height is 0.15m. Wave from source run-upped a shore.

When the wave propagates over the sloping bottom, its amplitude increases significantly due to the shallow-water effect. The crest of the wave continues to steepen and in the end this leads to a breaking of the wave.

Figure 5.5.1b shows the results of simulating the run-up and breaking of the surface wave along a sloping shore based on the Reynolds averaging method of the Navier-Stokes equations (RANS) of American researchers [Zhao *et al.*, 2004]. When propagating along on a sloping bottom, the surface wave has the following parameters: period T = 5s, wave height 0.128m.

Figure 5.5.1c shows the dynamics of the distortion of the profile of the surface gravity wave during propagation along on flat shore slope on the basis of our model.

The comparison allows us to note the following:

- for all three models the profile of the surface wave propagating on a sloping bottom begins to distort due to the impact of depth and the forefront slope of the wave crest becomes steeper;
- with a further decrease in the depth, the wave breaking is accompanied with a curl on the crest (Figure 5.5.1a and b); our model generates no curl because the two-phase air–water interaction was not considered; therefore, the point of breaking corresponds to a vertical position of the front slope of the wave crest (Figure 5.5.1c).

Next, we consider the experimental measurements of the surface wave profile during propagation along a sloping bottom. In the work of the French scientists [Kimmoun, Branger, 2007] experimental measurements of the breaking of wave are given taking into account the processes of sharpening, stiffening, and overturning.

The experiments were carried out in the Ecole Centrale basin in Marseilles (Figure 5.5.2a). The glass tank has a length of 17m and a width of 0.65m, water depth is 0.735m, the bottom slope is 1/15 (starting from 4m from the wave source). Cameras are located along the path of the wave. The wave period is $T = 1.3$s, the initial wave amplitude is $a = 0.07$m, and the wavelength is $\lambda =$ 2.5m. The wave breaks at a distance of 12.375m from the wave source.

Figure 5.5.2b shows photographs of the gradual distortion of the profile of the breaks wave. Twelve chambers are located on a section with a length of 3.27m near the boundary line (Figure 5.5.2a, 12 PIV windows). Images of the initial four stages of distortion of the wave profile are shown in Figure 5.5.2b. In the windows (w1) and (w2), the forefront of the wave is steepened, in the windows (w3) and (w4), a jet emerges from the wave crest to form a curl. Further, the curl falls to the front slope of the wave crest with the capture of the air mass, the formation of foam and bubbles.

Analyzing the experimental distortions of the surface wave profile, it can be noted:

- with decreasing depth, the forefront of the wave crest is steepened similarly to numerical results;
- Depending on taking into account the two-phase state of the propagation medium, the wave breaks by the formation of a curl on the crest or by a transition to the plumb condition of the forefront of the wave.

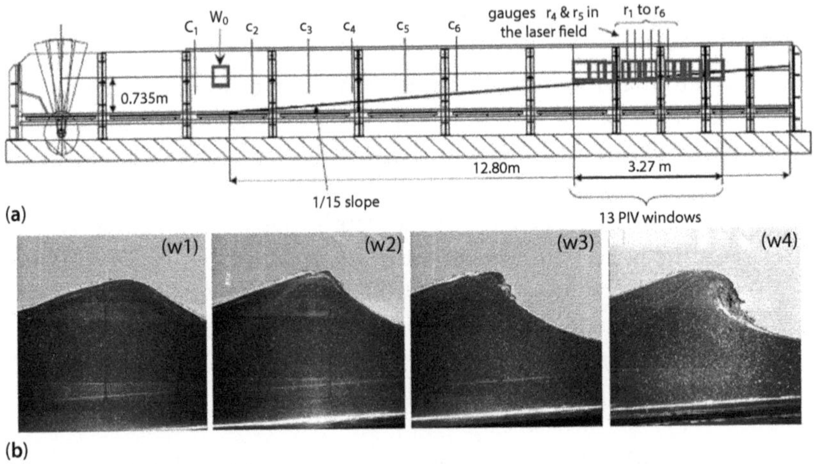

Figure 5.5.2 The experimental setup (a) and stage-by-stage distortions of the profile of the breaking surface wave propagating along the sloping bottom (b), the initial wave parameters: $\lambda = 2.5$m; H = 0.735m; $T = 1.3$s; $a = 0.07$m; distance of breaking 12.375m [Kimmoun, Branger, 2007].

As a result of the comparison of the results of a numerical simulation of the run-up of nonlinear surface gravity waves onto the coastal slopes within our consideration with numerical and experimental studies by other authors it can be noted that they have a good agreement on the initial stages of the distortion of the surface wave profile.

Conclusions

In the fifth chapter the questions of modeling the distribution of surface gravity waves based on computational solution of the Navier-Stokes equations were examined.

The questions of 2D computational modeling of run-up of nonlinear surface gravity waves based on Navier-Stokes equations were examined. The task statement was defined, the task geometry was presented, and the boundary and initial conditions were described. Using the physical process split method, a discrete model on a uniform grid was created. Mesh schemas of the original differential equations was received using the integral-interpolation method.

A discrete, finite-volume model of the hydrodynamic task to be investigated was developed, taking into account the cell occupancy factor. The discrete analogues of the equations were obtained to calculate the components of the velocity vector, the pressure fields, and the discrete analogues of the boundary conditions. However, the discrete equations for calculating the components of velocity and pressure fields are calculated on an implicit basis, and discrete equations to clarify the components of the velocity field are calculated by an explicit schema.

A study of the conservativeness of the discrete model was carried out. It was found that an integral law of impulse conservation is fulfilled for web analogy of the differential equation. An approximation error of a finite schema was found, which has the first order of time and second order in two spatial coordinates.

The over relaxation method was used to solve the grid equations, and the rationale for selecting the method and the characteristics of its work were presented. A 2DBayWaves program has been developed to calculate the 2D velocity and pressure field to model the run-up and fall the non-linear surface gravity wave on the shallow water coast, the structure and algorithm of the program were described.

The results of 2D computational modeling of the run-up of nonlinear surface gravity waves to the coast formations of shallow water are provided. Coastal slopes of different types were modeled, and graphs of the surface gravity wave profile changes which is running-up to the shore slopes. The process of run-up of a surface wave to the coast, rewinding it back, wave overlapping, steeping, and falling, are shown. In order to assess the depth of the run-up of the water masses the data was presented, which depends on the length and initial steepness of the nonlinear surface wave.

The impact of the bottom geometry on the wave run-up, the features of run-up waves were defined when the wave length and the depth of the shallow water change. The conditions for a complete flooding of the coastal zone in long shallow water areas with forward storm waves or tsunami waves have been modeled.

To test the adequacy of the model, the results of the modeling were compared to the existing computational and experimental data. The resulting comparison shows a good agreement with the data of the other authors, especially the initial stages of the distortion of the surface wave profile.

The results of the studies in this chapter have been published in the following main works: [Abbasov, 2015 Lenand], [Abbasov, Semenov, Tsarevsky, 2012], [Abbasov, 2013-1], [Abbasov, 2013-1].

6

Three-Dimensional Numerical Modeling of the Runup of Nonlinear Surface Gravity Waves

6.1 Statement of the Problem. Boundary and Initial Conditions

To describe three-dimensional surface gravity waves with allowance for viscosity, we use the three-dimensional Navier–Stokes equations, the continuity equation for an incompressible fluid, and the hydrodynamic pressure equation [Fletcher, 1991], [Samarsky, 1989]. Specifically,

- Navier-Stokes equations are:

$$\frac{\partial u}{\partial t}+u\frac{\partial u}{\partial x}+v\frac{\partial u}{\partial y}+w\frac{\partial u}{\partial z}=-\frac{1}{\rho}\frac{\partial P}{\partial x}+\mu\frac{\partial^2 u}{\partial x^2}+\mu\frac{\partial^2 u}{\partial y^2}+\eta\frac{\partial^2 u}{\partial z^2}, \quad (6.1.1)$$

$$\frac{\partial v}{\partial t}+u\frac{\partial v}{\partial x}+v\frac{\partial v}{\partial y}+w\frac{\partial v}{\partial z}=-\frac{1}{\rho}\frac{\partial P}{\partial y}+\mu\frac{\partial^2 v}{\partial x^2}+\mu\frac{\partial^2 v}{\partial y^2}+\eta\frac{\partial^2 v}{\partial z^2}, \quad (6.1.2)$$

$$\frac{\partial w}{\partial t}+u\frac{\partial w}{\partial x}+v\frac{\partial w}{\partial y}+w\frac{\partial w}{\partial z}=-\frac{1}{\rho}\frac{\partial P}{\partial z}+\mu\frac{\partial^2 w}{\partial x^2}+\mu\frac{\partial^2 w}{\partial y^2}+\eta\frac{\partial^2 w}{\partial z^2}+g. \tag{6.1.3}$$

- the incompressible continuity equation is:

$$\frac{\partial u}{\partial x}+\frac{\partial v}{\partial y}+\frac{\partial w}{\partial z}=0, \tag{6.1.4}$$

- the total hydrodynamic pressure equation of with allowance for the depth is:

$$P(x,y,z,t)=p(x,y,z,t)+\rho gz, \tag{6.1.5}$$

where u, v, – are the horizontal components and w – is the vertical component of the velocity $V(u,v, w)$ of water particles, ρ– is the water density, g – is the acceleration of gravity, μ, η – are the horizontal and vertical components of the turbulent exchange coefficient.

According to the geometry of the problem (Figure 6.1.1.), the Cartesian x axis is placed on the unperturbed water surface and is directed toward the coast, the y axis is directed along the coast, and the z axis is directed vertically upward. At the initial time, the water is still. At some distance from the coast, at the point $x = 0$, we set a perturbation in the form of a pressure pulse that varies harmonically while propagating away from the origin. Let us examine the resulting motion of the water mass toward the coast.

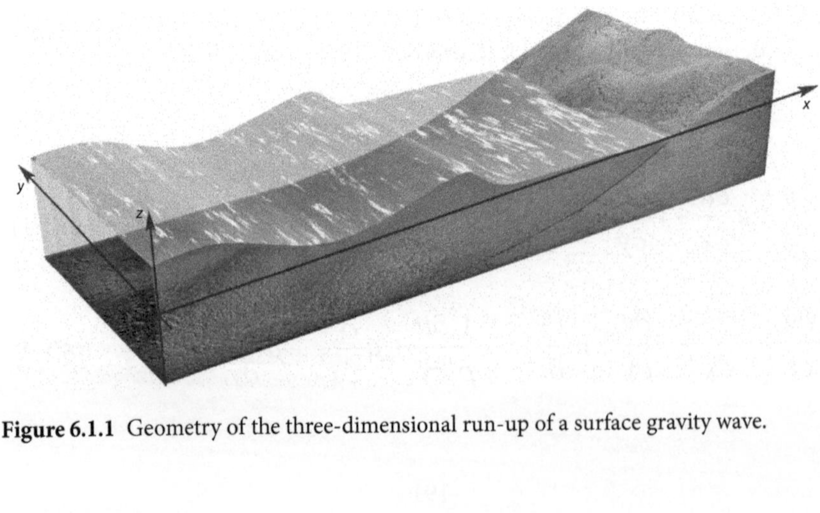

Figure 6.1.1 Geometry of the three-dimensional run-up of a surface gravity wave.

Boundary Conditions

At the bottom of the domain, we set the impermeability and slip conditions:

$$\begin{array}{l} V_n(x,y,z,t)=0,\ \dfrac{\partial}{\partial n}P(x,y,z,t)=0, \\ \dfrac{\partial}{\partial x}u(x,y,z,t)=0,\ \dfrac{\partial}{\partial y}v(x,y,z,t)=0,\ \dfrac{\partial}{\partial z}w(x,y,z,t)=0 \end{array}, \qquad (6.1.6)$$

$$\begin{array}{l} \rho\eta\dfrac{\partial}{\partial z}u(x,y,z,t)=-\tau_x(t),\ \rho\eta\dfrac{\partial}{\partial z}v(x,y,z,t)=-\tau_y(t), \\ \rho\mu\dfrac{\partial}{\partial x}w(x,y,z,t)=-\tau_z(t) \end{array};$$

- on the free surface of the fluid, the dynamic and kinematic conditions:

$$\begin{array}{l} \dfrac{\partial}{\partial n}u(x,y,z,t)=0,\ \dfrac{\partial}{\partial n}v(x,y,z,t)=0,\ \dfrac{\partial}{\partial n}w(x,y,z,t)=0 \\ w(x,y,z,t)=\dfrac{1}{g\rho}\dfrac{\partial P}{\partial t},\ \dfrac{\partial}{\partial n}P(x,y,z,t)=0 \end{array}, \qquad (6.1.7)$$

- at the lateral boundaries – the impermeability conditions:

$$\begin{array}{l} \dfrac{\partial}{\partial n}u(x,y,z,t)=0,\ \dfrac{\partial}{\partial n}v(x,y,z,t)=0, \\ \dfrac{\partial}{\partial n}w(x,y,z,t)=0,\ \dfrac{\partial}{\partial n}P(x,y,z,t)=\alpha \end{array}, \qquad (6.1.8)$$

where $\tau_x(t)$, $\tau_y(t)$, $\tau_z(t)$ – the components of the bottom shear stress, параметр α – is a parameter specified using the initial conditions.

Components of shear stress caused by bottom friction are computed using the formulas (according to the Van Dorn law [Fletcher, 1991], [Roach, 1980]):

$$\tau_x(t)=\rho C_p(|V|)u|V|,\quad \tau_y(t)=\rho C_p(|V|)v|V|,\quad \tau_z(t)=\rho C_p(|V|)w|V|, \qquad (6.1.8)$$

where

$$C_p(|V|) = \begin{cases} 0.0088; & |V| < 6.6m/s, \\ 0.0026; & |V| \geq 6.6m/s \end{cases}$$

– is a nondimensional coefficient.

The horizontal and vertical components of the turbulent exchange coefficient take the values ($0 < \mu \leq 1$ m²/s, $0 < \eta \leq 1$ m²/s).

At the initial time t=0, the fluid is still and the following conditions are satisfied:

$$P(x,y,z,0) = \rho gz,\ u(x,y,z,0) = 0,$$
$$v(x,y,z,0) = 0,\ w(x,y,z,0) = 0.$$

Before turning to the description of the discrete model, let us consider some features of the process of turbulent exchange in shallow water.

Features of Turbulent Exchange

During the dissipation of surface waves in the coastal zone, bottom friction is affected by turbulent mixing of the aqueous medium. However, in the shallow water, the gradient of horizontal turbulent viscosity is insignificant in comparison with the vortex viscosity in the vertical direction. Usually the coefficient of horizontal turbulent viscosity is assumed to be constant or a priori determined. For distances of the order of 100 ÷ 1000 km, the value of the coefficient of horizontal turbulent viscosity μ can be taken to be of the order of $10^2 \div 10^3$ m²/s [Sukhinov et al., 2011].

Several methods are used to determine the coefficient of vertical turbulent viscosity η. In most hydrodynamic models, the coefficient of vertical turbulent mixing does not always agree with experimental measurements.

The constancy of the coefficient of vertical turbulent exchange cannot give a physically justified picture of hydrodynamic processes. This is due to the fact that at different depths of shallow water areas the coefficient of vertical turbulent exchange can differ by tens of times. Turbulent exchange on the vertex can arise either as a result of shear flows, or due to unstable stratification.

In the coastal zone, a shear flow can occur due to wind, tidal, surface and baroclinic currents. Unstable stratification arises from cooling, evaporation, and advective processes. Consequently, the constant coefficient of vertical turbulent exchange does not allow us to obtain a three-dimensional velocity vector of the aquatic medium with the required accuracy.

In modern numerical hydrodynamic models, the coefficient of vertical turbulent exchange is used as an adjustable parameter, by means of which it is possible to obtain sufficiently good agreement between the results obtained and the experimental observations. Its value is often chosen also from considerations of computational stability in calculations for longer times and may exceed by orders of magnitude physically reasonable estimates of the coefficient for a real marine environment.

Therefore, when choosing the values of the coefficient of vertical turbulent exchange, it is necessary first of all to be guided by the data of field experiments. It should be noted that when *splitting the shallow water equations*, we also introduced the coefficient of artificial viscosity μ for stability purposes.

[Sukhinov et al., 2011] presents the results of expeditions to measure data on the state of hydrophysical and hydrochemical parameters conducted from 2005 to 2009 in Taganrog Bay and in the eastern part of the Azov Sea. During these expeditions, data were obtained on the instantaneous values and pulsations of three-dimensional velocity vectors as a function of depth in many tens of shallow water points using the ADCP (Acoustic Doppler Current Profiler) probe WHS600 Sentinel. Experiments were performed on the basis of several approaches for calculating the coefficient of turbulent exchange along the vertical for all points, as a result of which the distributions of vertical turbulent exchange coefficients for the Sea of Azov were obtained.

6.2 Construction of a Discrete Model

The computational domain in space is a cube. To implement the discrete mathematical model numerically, we introduce the uniform grid:

$$\varpi_h = \Big\{x_i = ih_x, y_j = jh_y, z_k = kh_z; i = \overline{0..N_x}, j = \overline{0..N_z};$$
$$k = \overline{0..N_k}; N_x h_x = l_x, N_y h_y = l_y, N_z h_z = l_z\Big\};$$

where i, j, k – are the indices in x, y, z directions, respectively; h_x, h_y, h_z – are the mesh sizes in space; N_x, N_y, N_z – are the numbers of nodes in respective spatial directions; l_x, l_y, l_z – are the spatial sizes of the domain.

We approximate the equations (6.1.1), (6.1.2) and (6.1.3) by the time variable and we introduce an intermediate layer $n + \sigma$:

$$\frac{u^{n+1} - u^{n+\sigma} + u^{n+\sigma} - u^{n}}{\tau} + u\frac{\partial u}{\partial x} + v\frac{\partial u}{\partial y} + w\frac{\partial u}{\partial z}$$

$$= -\frac{1}{\rho}\frac{\partial P}{\partial x} + \mu\frac{\partial^2 u}{\partial x^2} + \mu\frac{\partial^2 u}{\partial y^2} + \eta\frac{\partial^2 u}{\partial z^2}, \tag{6.2.1}$$

$$\frac{v^{n+1} - v^{n+\sigma} + v^{n+\sigma} - v^{n}}{\tau} + u\frac{\partial v}{\partial x} + v\frac{\partial v}{\partial y} + w\frac{\partial v}{\partial z}$$

$$= -\frac{1}{\rho}\frac{\partial P}{\partial y} + \mu\frac{\partial^2 v}{\partial x^2} + \mu\frac{\partial^2 v}{\partial y^2} + \eta\frac{\partial^2 v}{\partial z^2}, \tag{6.2.2}$$

$$\frac{w^{n+1} - w^{n+\sigma} + w^{n+\sigma} - w^{n}}{\tau} + u\frac{\partial w}{\partial x} + v\frac{\partial w}{\partial y} + w\frac{\partial w}{\partial z}$$

$$= -\frac{1}{\rho}\frac{\partial P}{\partial z} + \mu\frac{\partial^2 w}{\partial x^2} + \mu\frac{\partial^2 w}{\partial y^2} + \eta\frac{\partial^2 w}{\partial z^2} + g. \tag{6.2.3}$$

We apply the *method of splitting* the physical processes into equations (6.2.1), (6.2.2) и (6.2.3)

$$\frac{u^{n+\sigma} - u^{n}}{\tau} + u\frac{\partial u}{\partial x} + v\frac{\partial u}{\partial y} + w\frac{\partial u}{\partial z} = \mu\frac{\partial^2 u}{\partial x^2} + \mu\frac{\partial^2 u}{\partial y^2} + \eta\frac{\partial^2 u}{\partial z^2}, \tag{6.2.4}$$

$$\frac{v^{n+\sigma} - v^{n}}{\tau} + u\frac{\partial v}{\partial x} + v\frac{\partial v}{\partial y} + w\frac{\partial v}{\partial z} = \mu\frac{\partial^2 v}{\partial x^2} + \mu\frac{\partial^2 v}{\partial y^2} + \eta\frac{\partial^2 v}{\partial z^2}, \tag{6.2.5}$$

$$\frac{w^{n+\sigma} - w^{n}}{\tau} + u\frac{\partial w}{\partial x} + v\frac{\partial w}{\partial y} + w\frac{\partial w}{\partial z} = \mu\frac{\partial^2 w}{\partial x^2} + \mu\frac{\partial^2 w}{\partial y^2} + \eta\frac{\partial^2 w}{\partial z^2} + g, \tag{6.2.6}$$

$$\frac{u^{n+1} - u^{n+\sigma}}{\tau} = -\frac{1}{\rho}\frac{\partial P}{\partial x}, \tag{6.2.7}$$

$$\frac{v^{n+1} - v^{n+\sigma}}{\tau} = -\frac{1}{\rho}\frac{\partial P}{\partial y}, \tag{6.2.8}$$

$$\frac{w^{n+1}-w^{n+\sigma}}{\tau}=-\frac{1}{\rho}\frac{\partial P}{\partial z}. \qquad (6.2.9)$$

After differentiating equations (6.2.7), (6.2.8), (6.2.9), taking into account the continuity equation for the pressure, we obtain:

$$\frac{\partial^2 P}{\partial x^2}+\frac{\partial^2 P}{\partial y^2}+\frac{\partial^2 P}{\partial z^2}=\frac{\rho}{\tau}\left(\frac{\partial u^{n+\sigma}}{\partial x}+\frac{\partial v^{n+\sigma}}{\partial y}+\frac{\partial w^{n+\sigma}}{\partial z}\right). \qquad (6.2.10)$$

According to the method of splitting by physical processes, the calculation of the wave hydrodynamics problem is carried out in three stages:

- in the first stage, compute the velocity field by applying the equations (6.2.4), (6.2.5), (6.2.6);
- the second stage calculates the pressure from equation (6.2.10);
- i the third stage, on the basis of equations (6.2.7), (6.2.8), (6.2.9) the velocity field is determined from the pressure.
- The boundary conditions for the first stage are:
- at the bottom of the domain:

$$\frac{\partial}{\partial x}u(x,y,z,t)=0,\ \frac{\partial}{\partial y}v(x,y,z,t)=0,\ \frac{\partial}{\partial z}w(x,y,z,t)=0, \qquad (6.2.11)$$

$$\rho\eta\frac{\partial}{\partial z}u(x,y,z,t)=-\tau_x(t),\ \rho\eta\frac{\partial}{\partial z}v(x,y,z,t)$$
$$=-\tau_y(t),\ \rho\mu\frac{\partial}{\partial x}w(x,y,z,t)$$
$$=-\tau_z(t);$$

- on the free surface of the fluid and the lateral boundaries:

$$\frac{\partial}{\partial n}u(x,y,z,t)=0,\ \frac{\partial}{\partial n}v(x,y,z,t)=0,\ \frac{\partial}{\partial n}w(x,y,z,t)=0. \qquad (6.2.12)$$

The boundary conditions for the second stage are:

- at the bottom of the domain:

$$V_n(x,y,z,t)=0,\ \frac{\partial}{\partial n}P(x,y,z,t)=0,$$
$$\frac{\partial}{\partial x}u(x,y,z,t)=0,\ \frac{\partial}{\partial y}v(x,y,z,t)=0, \tag{6.2.13}$$
$$\frac{\partial}{\partial z}w(x,y,z,t)=0,$$

- on the free surface of the fluid:

$$\frac{\partial}{\partial n}u(x,y,z,t)=0,\ \frac{\partial}{\partial n}v(x,y,z,t)=0,\ \frac{\partial}{\partial n}w(x,y,z,t)=0.$$
$$w(x,y,z,t)=\frac{1}{g\rho}\frac{\partial P}{\partial t},\ \frac{\partial}{\partial n}P(x,y,z,t)=0, \tag{6.2.14}$$

on the lateral boundaries:

$$\frac{\partial}{\partial n}u(x,y,z,t)=0,\ \frac{\partial}{\partial n}v(x,y,z,t)=0,\ \frac{\partial}{\partial n}w(x,y,z,t)=0.$$

$$V_n(x,y,z,t)=0,\ \frac{\partial}{\partial n}P(x,y,z,t)=\alpha. \tag{6.2.15}$$

The boundary condition for the third stage is:

$$\frac{\partial}{\partial n}P(x,y,z,t)=0. \tag{6.2.16}$$

We approximate the equations (6.2.4) - (6.2.9), (6.2.10) by the space variables. For this we use the integro-interpolation method, integrate equation (6.2.4) over the domain Θ:

$$x\in\left[(i-1/2)h_x,\ (i+1/2)h_x\right];$$

$$y\in\left[(j-1/2)h_y,\ (j+1/2)h_y\right];$$

$$z\in\left[(k-1/2)h_z,\ (k+1/2)h_z\right].$$

$$\begin{aligned}&\iiint\limits_{\Theta}\frac{u^{n+\sigma}-u^{n}}{\tau}dxdydz+\iiint\limits_{\Theta}u\frac{\partial u}{\partial x}dxdydz+\iiint\limits_{\Theta}v\frac{\partial u}{\partial y}dxdydz\\&+\iiint\limits_{\Theta}w\frac{\partial u}{\partial z}dxdydz=\iiint\limits_{\Theta}\mu\frac{\partial^{2}u}{\partial x^{2}}dxdydz+\iiint\limits_{\Theta}\mu\frac{\partial^{2}u}{\partial y^{2}}dxdydz\\&+\iiint\limits_{\Theta}\eta\frac{\partial^{2}u}{\partial z^{2}}dxdydz.\end{aligned}\tag{6.2.17}$$

We calculate each integral separately taking into account the following notation:

$$u_{i,j,k}=\frac{1}{h_xh_yh_z}\iiint\limits_{\Theta}udxdydz;\ v_{i,j,k}=\frac{1}{h_xh_yh_z}\iiint\limits_{\Theta}vdxdydz;\tag{6.2.18}$$

$$w_{i,j,k}=\frac{1}{h_xh_yh_z}\iiint\limits_{\Theta}wdxdydz.$$

The first integral of (6.2.17) takes the form:

$$\iiint\limits_{\Theta}\frac{u^{n+\sigma}-u^{n}}{\tau}dxdydz=h_xh_yh_z\frac{u_{i,j,k}^{n+\sigma}-u_{i,j,k}^{n}}{\tau}.\tag{6.2.19}$$

Compute the second integral of expression (6.2.17):

$$\begin{aligned}&\iiint\limits_{\Theta}u\frac{\partial u}{\partial x}dxdydz\\&=\int\limits_{z_{k-1/2}}^{z_{k+1/2}}dz\int\limits_{y_{j-1/2}}^{y_{j+1/2}}dy\int\limits_{x_{i-1/2}}^{x_{i+1/2}}u\frac{\partial u}{\partial x}dx=\int\limits_{z_{k-1/2}}^{z_{k+1/2}}dz\int\limits_{y_{j-1/2}}^{y_{j+1/2}}dy\int\limits_{x_{i-1/2}}^{x_{i}}u\frac{\partial u}{\partial x}dx\\&\quad+\int\limits_{z_{k-1/2}}^{z_{k+1/2}}dz\int\limits_{y_{j-1/2}}^{y_{j+1/2}}dy\int\limits_{x_{i}}^{x_{i+1/2}}u\frac{\partial u}{\partial x}dx\\&=u_{i-1/2,j,k}\int\limits_{z_{k-1/2}}^{z_{k+1/2}}dz\int\limits_{y_{j-1/2}}^{y_{j+1/2}}u\Big|_{x_{i-1/2}}^{x_{i}}dy+u_{i+1/2,j,k}\int\limits_{z_{k-1/2}}^{z_{k+1/2}}dz\int\limits_{y_{j-1/2}}^{y_{j+1/2}}u\Big|_{x_{i}}^{x_{i+1/2}}dy\\&=u_{i-1/2,j,k}h_zh_y(u_{i,j,k}-u_{i-1,j,k})+u_{i+1/2,j,k}h_zh_y(u_{i+1,j,k}-u_{i,j,k})\end{aligned}\tag{6.2.20}$$

In the final form for the second integral we obtain:

$$\iiint_{\Theta} u\frac{\partial u}{\partial x}dxdydz = u_{i-1/2,j,k}h_z h_y(u_{i,j,k} - u_{i-1,j,k}) + u_{i+1/2,j,k}h_z h_y(u_{i+1,j,k} - u_{i,j,k}). \tag{6.2.21}$$

Similarly, for the third

$$\begin{aligned}&\iiint_{\Theta} v\frac{\partial u}{\partial y}dxdydz \\ &= v_{i,j-1/2,k}h_z h_x(u_{i,j,k} - u_{i,j-1,k}) + v_{i,j+1/2,k}h_z h_x(u_{i,j+1,k} - u_{i,j,k}).\end{aligned} \tag{6.2.22}$$

and the fourth integrals, we obtain:

$$\begin{aligned}&\iiint_{\Theta} w\frac{\partial u}{\partial z}dxdydz \\ &= w_{i,j,k-1/2}h_x h_y(u_{i,j,k} - u_{i,j,k-1}) + w_{i,j,k+1/2}h_x h_y(u_{i,j,k+1} - u_{i,j,k}).\end{aligned} \tag{6.2.23}$$

Let us turn to the integrals on the right side of the equation (6.2.17)

$$\iiint_{\Theta} \mu\frac{\partial^2 u}{\partial x^2}dxdydz = \int_{z_{k-1/2}}^{z_{k+1/2}} dz \int_{y_{j-1/2}}^{y_{j+1/2}} dy \int_{x_{i-1/2}}^{x_{i+1/2}} \left(\mu\frac{\partial^2 u}{\partial x^2}\right)dx = \int_{z_{k-1/2}}^{z_{k+1/2}} \int_{y_{j-1/2}}^{y_{j+1/2}} \mu\frac{\partial u}{\partial x}\bigg|_{x_{i-1/2}}^{x_{i+1/2}} dy. \tag{6.2.24}$$

We integrate the derivative on the interval $x \in \left[x_i, x_{i+1}\right]$, then we obtain:

$$\int_{x_i}^{x_{i+1}} \mu\frac{\partial u}{\partial x}dx = \mu_{i+1/2}\frac{u_{i+1} - u_i}{h_x}.$$

In the final form, expression (6.2.24) takes the form:

$$\iiint_{\Theta} \mu\frac{\partial^2 u}{\partial x^2}dxdydz = h_z h_y\left(\mu_{i+1/2,j,k}\frac{u_{i+1,j,k} - u_{i,j,k}}{h_x} - \mu_{i-1/2,j,k}\frac{u_{i,j,k} - u_{i-1,j,k}}{h_x}\right). \tag{6.2.25}$$

Similarly, one can obtain a solution for the second

$$\iiint\limits_{\Theta} \mu \frac{\partial^2 u}{\partial y^2} dxdydz = h_x h_z \left(\mu_{i,j+1/2,k} \frac{u_{i,j+1,k} - u_{i,j,k}}{h_y} - \mu_{i,j-1/2,k} \frac{u_{i,j,k} - u_{i,j-1,k}}{h_y} \right). \tag{6.2.26}$$

and third integrals on the right side of the expression (6.2.17):

$$\iiint\limits_{\Theta} \eta \frac{\partial^2 u}{\partial z^2} dxdydz = h_x h_y \left(\eta_{i,j,k+1/2} \frac{u_{i,j,k+1} - u_{i,j,k}}{h_z} - \eta_{i,j,k-1/2} \frac{u_{i,j,k} - u_{i,j,k-1}}{h_z} \right). \tag{6.2.27}$$

We substitute the obtained solutions of the integrals in the original expression (6.2.17) (taking into account the division by the cell volume $h_x h_y h_z$):

$$\frac{u^{n+\sigma}_{i,j,k} - u^n_{i,j,k}}{\tau} + \frac{u_{i-1/2,j,k}\left(u_{i,j,k} - u_{i-1,j,k}\right)}{2h_x} + \frac{u_{i+1/2,j,k}\left(u_{i+1,j,k} - u_{i,j,k}\right)}{2h_x} +$$

$$+ \frac{v_{i,j-1/2,k}\left(u_{i,j,k} - u_{i,j-1,k}\right)}{2h_y} + \frac{v_{i,j+1/2,k}\left(u_{i,j+1,k} - u_{i,j,k}\right)}{2h_y} +$$

$$+ \frac{w_{i,j,k-1/2}\left(u_{i,j,k} - u_{i,j,k-1}\right)}{2h_z} + \frac{w_{i,j,k+1/2}\left(u_{i,j,k+1} - u_{i,j,k}\right)}{2h_z} =$$

$$= \left(\mu_{i+1/2,j,k} \frac{u_{i+1,j,k} - u_{i,j,k}}{h_x^2} - \mu_{i-1/2,j,k} \frac{u_{i,j,k} - u_{i-1,j,k}}{h_x^2} \right) +$$

$$+ \left(\mu_{i,j+1/2,k} \frac{u_{i,j+1,k} - u_{i,j,k}}{h_y^2} - \mu_{i,j-1/2,k} \frac{u_{i,j,k} - u_{i,j-1,k}}{h_y^2} \right) +$$

$$+ \left(\eta_{i,j,k+1/2} \frac{u_{i,j,k+1} - u_{i,j,k}}{h_z^2} - \eta_{i,j,k-1/2} \frac{u_{i,j,k} - u_{i,j,k-1}}{h_z^2} \right). \tag{6.2.28}$$

Expression (6.2.28) is a finite-difference scheme of equation (6.2.17). Similarly, we can obtain a finite-difference scheme of equation (6.2.5):

$$\frac{v_{i,j,k}^{n+\sigma}-v_{i,j,k}^{n}}{\tau}+\frac{u_{i-1/2,j,k}\left(v_{i,j,k}-v_{i-1,j,k}\right)}{2h_x}+\frac{u_{i+1/2,j,k}\left(v_{i+1,j,k}-v_{i,j,k}\right)}{2h_x}+$$

$$+\frac{v_{i,j-1/2,k}\left(v_{i,j,k}-v_{i,j-1,k}\right)}{2h_y}+\frac{v_{i,j+1/2,k}\left(v_{i,j+1,k}-v_{i,j,k}\right)}{2h_y}+$$

$$+\frac{w_{i,j,k-1/2}\left(v_{i,j,k}-v_{i,j,k-1}\right)}{2h_z}+\frac{w_{i,j,k+1/2}\left(v_{i,j,k+1}-v_{i,j,k}\right)}{2h_z}=$$

$$=\left(\mu_{i+1/2,j,k}\frac{v_{i+1,j,k}-u_{i,j,k}}{h_x^2}-\mu_{i-1/2,j,k}\frac{v_{i,j,k}-v_{i-1,j,k}}{h_x^2}\right)+$$

$$+\left(\mu_{i,j+1/2,k}\frac{v_{i,j+1,k}-v_{i,j,k}}{h_y^2}-\mu_{i,j-1/2,k}\frac{v_{i,j,k}-v_{i,j-1,k}}{h_y^2}\right)+$$

$$+\left(\eta_{i,j,k+1/2}\frac{v_{i,j,k+1}-v_{i,j,k}}{h_z^2}-\eta_{i,j,k-1/2}\frac{v_{i,j,k}-v_{i,j,k-1}}{h_z^2}\right). \tag{6.2.29}$$

Also, for the equation (6.2.6), we can obtain a finite-difference scheme:

$$\frac{w_{i,j,k}^{n+\sigma}-w_{i,j,k}^{n}}{\tau}+\frac{u_{i-1/2,j,k}\left(w_{i,j,k}-w_{i-1,j,k}\right)}{2h_x}+\frac{u_{i+1/2,j,k}\left(w_{i+1,j,k}-w_{i,j,k}\right)}{2h_x}+$$

$$+\frac{v_{i,j-1/2,k}\left(w_{i,j,k}-w_{i,j-1,k}\right)}{2h_y}+\frac{v_{i,j+1/2,k}\left(w_{i,j+1,k}-w_{i,j,k}\right)}{2h_y}+$$

$$+\frac{w_{i,j,k-1/2}\left(w_{i,j,k}-w_{i,j,k-1}\right)}{2h_z}+\frac{w_{i,j,k+1/2}\left(w_{i,j,k+1}-w_{i,j,k}\right)}{2h_z}=$$

$$=\left(\mu_{i+1/2,j,k}\frac{w_{i+1,j,k}-w_{i,j,k}}{h_x^2}-\mu_{i-1/2,j,k}\frac{w_{i,j,k}-w_{i-1,j,k}}{h_x^2}\right)+$$

$$+\left(\mu_{i,j+1/2,k}\frac{w_{i,j+1,k}-w_{i,j,k}}{h_y^2}-\mu_{i,j-1/2,k}\frac{w_{i,j,k}-w_{i,j-1,k}}{h_y^2}\right)+$$

$$+\left(\eta_{i,j,k+1/2}\frac{w_{i,j,k+1}-w_{i,j,k}}{h_z^2}-\eta_{i,j,k-1/2}\frac{w_{i,j,k}-w_{i,j,k-1}}{h_z^2}\right)+g\,. \quad (6.2.30)$$

Next we proceed to approximate the diffusion and convection operators on a time variable applying schemes with weights ($\sigma = 0.5$).

We integrate the equation (6.2.10) over the domain Θ:

$$x \in \left[(i-1/2)h_x,\ (i+1/2)h_x\right];$$

$$y \in \left[(j-1/2)h_y,\ (j+1/2)h_y\right];$$

$$z \in \left[(k-1/2)h_z,\ (k+1/2)h_z\right].$$

then we get:

$$\iiint_\Theta \frac{\partial^2 P}{\partial x^2}dxdydz+\iiint_\Theta \frac{\partial^2 P}{\partial y^2}dxdydz+\iiint_\Theta \frac{\partial^2 P}{\partial z^2}dxdydz$$
$$=\frac{\rho}{\tau}\left(\iiint_\Theta \frac{\partial u^{n+\sigma}}{\partial x}dxdydz+\iiint_\Theta \frac{\partial v^{n+\sigma}}{\partial y}dxdydz+\iiint_\Theta \frac{\partial w^{n+\sigma}}{\partial z}dxdydz\right). \quad (6.2.31)$$

Consider the first integral of expression (6.2.31)

$$\iiint_\Theta \frac{\partial^2 P}{\partial x^2}dxdydz=\int_{z_{k-1/2}}^{z_{k+1/2}}dz\int_{y_{j-1/2}}^{y_{j+1/2}}dy\int_{x_{i-1/2}}^{x_{i+1/2}}\frac{\partial^2 P}{\partial x^2}dx$$
$$=\int_{z_{k-1/2}}^{z_{k+1/2}}\int_{y_{j-1/2}}^{y_{j+1/2}}\left.\frac{\partial P}{\partial x}\right|_{x_{i-1/2}}^{x_{i+1/2}}dy \quad (6.2.32)$$
$$=h_y h_z\left(\frac{P_{i+1,j,k}-P_{i,j,k}}{h_x}-\frac{P_{i,j,k}-P_{i-1,j,k}}{h_x}\right).$$

Similarly, one can obtain for the second

$$\iiint_{\Theta} \frac{\partial^2 P}{\partial y^2} dxdydz = h_x h_z \left(\frac{P_{i,j+1,k} - P_{i,j,k}}{h_y} - \frac{P_{i,j,k} - P_{i,j-1,k}}{h_y} \right), \quad (6.2.33)$$

and third integrals on the left side:

$$\iiint_{\Theta} \frac{\partial^2 P}{\partial z^2} dxdydz = h_x h_y \left(\frac{P_{i,j,k+1} - P_{i,j,k}}{h_z} - \frac{P_{i,j,k} - P_{i,j,k-1}}{h_z} \right). \quad (6.2.34)$$

Consider the first integral on the right side of the expression (6.2.31)

$$\iiint_{\Theta} \frac{\partial u^{n+\sigma}}{\partial x} dxdydz = \int_{z_{k-1/2}}^{z_{k+1/2}} dz \int_{y_{j-1/2}}^{y_{j+1/2}} dy \int_{x_{i-1/2}}^{x_{i+1/2}} \frac{\partial u^{n+\sigma}}{\partial x} dx = \int_{z_{k-1/2}}^{z_{k+1/2}} \int_{y_{j-1/2}}^{y_{j+1/2}} (u^{n+\sigma})\Big|_{x_{i-1/2}}^{x_{i+1/2}} dy$$
$$= h_y h_z (u^{n+\sigma}_{i+1/2,j,k} - u^{n+\sigma}_{i-1/2,j,k}). \quad (6.2.35)$$

Similarly, one can obtain for the second

$$\iiint_{\Theta} \frac{\partial v^{n+\sigma}}{\partial y} dxdydz = h_x h_z (v^{n+\sigma}_{i,j+1/2,k} - v^{n+\sigma}_{i,j-1/2,k}), \quad (6.2.36)$$

and third integrals on the right side of the expression (6.2.31):

$$\iiint_{\Theta} \frac{\partial w^{n+\sigma}}{\partial z} dxdydz = h_x h_y (w^{n+\sigma}_{i,j,k+1/2} - w^{n+\sigma}_{i,j,k-1/2}). \quad (6.2.37)$$

We substitute the solutions obtained into the original equation (6.2.31) (with allowance for the division by $h_x h_y h_z$):

$$\frac{P_{i+1,j,k} - 2P_{i,j,k} + P_{i-1,j,k}}{h_x^2} + \frac{P_{i,j+1,k} - 2P_{i,j,k} + P_{i,j-1,k}}{h_y^2} + \frac{P_{i,j,k+1} - 2P_{i,j,k} + P_{i,j,k-1}}{h_z^2}$$
$$= \frac{\rho}{\tau} \left(\frac{u^{n+\sigma}_{i+1/2,j,k} - u^{n+\sigma}_{i-1/2,j,k}}{h_x} + \frac{v^{n+\sigma}_{i,j+1/2,k} - v^{n+\sigma}_{i,j-1/2,k}}{h_y} + \frac{w^{n+\sigma}_{i,j,k+1/2} - w^{n+\sigma}_{i,j,k-1/2}}{h_z} \right). \quad (6.2.38)$$

Expression (6.2.38) is a finite-difference scheme of equation (6.2.10).

We now turn to the equations for calculating the velocity components. We integrate the equation (6.2.7) over the domain Θ:

$$x \in \left[(i-1/2)h_x,\ (i+1/2)h_x\right];$$

$$y \in \left[(j-1/2)h_y,\ (j+1/2)h_y\right];$$

$$z \in \left[(k-1/2)h_z,\ (k+1/2)h_z\right].$$

$$\iiint\limits_{\Theta} \frac{u^{n+1} - u^{n+\sigma}}{\tau} dxdydz = -\frac{1}{\rho} \iiint\limits_{\Theta} \frac{\partial P}{\partial x} dxdydz . \qquad (6.2.39)$$

For the integral on the left side obtain:

$$\iiint\limits_{\Theta} \frac{u^{n+1} - u^{n+\sigma}}{\tau} dxdydz = \frac{u_{i,j,k}^{n+1} - u_{i,j,k}^{n+\sigma}}{\tau} h_x h_y h_z . \qquad (6.2.40)$$

For the integral on the right side:

$$\begin{aligned} \frac{1}{\rho} \iiint\limits_{\Theta} \frac{\partial P}{\partial x} dxdydz &= \frac{1}{\rho} \int\limits_{z_{k-1/2}}^{z_{k+1/2}} dz \int\limits_{y_{j-1/2}}^{y_{j+1/2}} dy \int\limits_{x_{i-1/2}}^{x_{i+1/2}} \frac{\partial P}{\partial x} dx \\ &= \frac{1}{\rho} \int\limits_{z_{k-1/2}}^{z_{k+1/2}} dz \int\limits_{y_{j-1/2}}^{y_{j+1/2}} P\Big|_{x_{i-1/2}}^{x_{i+1/2}} dy = \frac{h_y h_z}{\rho} \left(P_{i+1/2,j,k} - P_{i-1/2,j,k} \right). \end{aligned} \qquad (6.2.41)$$

To obtain the finite-difference scheme of equation (6.2.7), we substitute the solutions obtained in the original expression (6.2.39) (with allowance for the division by $h_x h_y h_z$):

$$\frac{u_{i,j,k}^{n+1} - u_{i,j,k}^{n+\sigma}}{\tau} = -\frac{1}{\rho} \left(\frac{P_{i+1/2,j,k} - P_{i-1/2,j,k}}{h_x} \right). \qquad (6.2.42)$$

After similar transformations, it is possible to obtain a finite-difference scheme of equation (6.2.8):

$$\frac{v_{i,j,k}^{n+1}-v_{i,j,k}^{n+\sigma}}{\tau}=-\frac{1}{\rho}\left(\frac{P_{i,j+1/2,k}-P_{i,j-1/2,k}}{h_y}\right). \tag{6.2.43}$$

The finite-difference scheme of equation (6.2.9) will have the form:

$$\frac{w_{i,j,k}^{n+1}-w_{i,j,k}^{n+\sigma}}{\tau}=-\frac{1}{\rho}\left(\frac{P_{i,j,k+1/2}-P_{i,j,k+1/2}}{h_z}\right). \tag{6.2.44}$$

We have obtained finite-difference schemes for the system of equations (6.2.4) ÷ (6.2.10).

6.3 The Construction of a Discrete Finite-Volume Model

The problem of wave hydrodynamics under consideration consists of three subtasks. The first subtask includes the system of equations (6.2.4), (6.2.5) and (6.2.6) for calculating the velocity field without taking the pressure into account:

$$\frac{u^{n+\sigma}-u^{n}}{\tau}+u\frac{\partial u}{\partial x}+v\frac{\partial u}{\partial y}+w\frac{\partial u}{\partial z}=\mu\frac{\partial^2 u}{\partial x^2}+\mu\frac{\partial^2 u}{\partial y^2}+\eta\frac{\partial^2 u}{\partial z^2}, \tag{6.2.4}$$

$$\frac{v^{n+\sigma}-v^{n}}{\tau}+u\frac{\partial v}{\partial x}+v\frac{\partial v}{\partial y}+w\frac{\partial v}{\partial z}=\mu\frac{\partial^2 v}{\partial x^2}+\mu\frac{\partial^2 v}{\partial y^2}+\eta\frac{\partial^2 v}{\partial z^2}, \tag{6.2.5}$$

$$\frac{w^{n+\sigma}-w^{n}}{\tau}+u\frac{\partial w}{\partial x}+v\frac{\partial w}{\partial y}+w\frac{\partial w}{\partial z}=\mu\frac{\partial^2 w}{\partial x^2}+\mu\frac{\partial^2 w}{\partial y^2}+\eta\frac{\partial^2 w}{\partial z^2}+g, \tag{6.2.6}$$

The boundary conditions are the following:

- at the bottom of the domain:

$$\frac{\partial}{\partial x}u(x,y,z,t)=0,\ \frac{\partial}{\partial y}v(x,y,z,t)=0,\ \frac{\partial}{\partial z}w(x,y,z,t)=0, \tag{6.3.1}$$

$$\rho\eta\frac{\partial}{\partial y}u(x,y,z,t)=-\tau_{xz}(t),\ \rho\eta\frac{\partial}{\partial x}v(x,y,z,t)$$
$$=-\tau_{yz}(t),\ \rho\mu\frac{\partial}{\partial z}w(x,y,z,t)=-\tau_{xy}(t);$$

- on a free surface and lateral boundaries:

$$\frac{\partial}{\partial n}u(x,y,z,t)=0,\ \frac{\partial}{\partial n}v(x,y,z,t)=0,\ \frac{\partial}{\partial n}w(x,y,z,t)=0. \quad (6.3.2)$$

The second subtask includes equation (6.2.10) for calculating pressure without taking into account the influence of gravity:

$$\frac{\partial^2 P}{\partial x^2}+\frac{\partial^2 P}{\partial y^2}+\frac{\partial^2 P}{\partial z^2}=\frac{\rho}{\tau}\left(\frac{\partial u^{n+\sigma}}{\partial x}+\frac{\partial v^{n+\sigma}}{\partial y}+\frac{\partial w^{n+\sigma}}{\partial z}\right). \quad (6.2.10)$$

The boundary conditions are:

- at the bottom of the domain:

$$\begin{aligned}&V_n(x,y,z,t)=0,\ \frac{\partial}{\partial n}P(x,y,z,t)=0,\\&\frac{\partial}{\partial x}u(x,y,z,t)=0,\ \frac{\partial}{\partial y}v(x,y,z,t)=0,\ \frac{\partial}{\partial z}w(x,y,z,t)=0,\end{aligned} \quad (6.3.3)$$

- on the free surface of the liquid:

$$\begin{aligned}&\frac{\partial}{\partial n}u(x,y,z,t)=0,\ \frac{\partial}{\partial n}v(x,y,z,t)=0,\ \frac{\partial}{\partial n}w(x,y,z,t)=0,\\&w(x,y,z,t)=\frac{1}{g\rho}\frac{\partial P}{\partial t},\ \frac{\partial}{\partial n}P(x,y,z,t)=0;\end{aligned} \quad (6.3.4)$$

- at the lateral border:

$$\begin{aligned}&\frac{\partial}{\partial n}u(x,y,z,t)=0,\ \frac{\partial}{\partial n}v(x,y,z,t)=0,\ \frac{\partial}{\partial n}w(x,y,z,t)=0,\\&V_n(x,y,z,t)=0,\ \frac{\partial}{\partial n}P(x,y,z,t)=\alpha.\end{aligned} \quad (6.3.5)$$

The third subtask consists of the system of equations (6.2.7), (6.2.8) and (6.2.9) for calculating the velocity field with allowance for the effect of pressure:

$$\frac{u^{n+1} - u^{n+\sigma}}{\tau} = -\frac{1}{\rho}\frac{\partial P}{\partial x}, \tag{6.2.7}$$

$$\frac{v^{n+1} - v^{n+\sigma}}{\tau} = -\frac{1}{\rho}\frac{\partial P}{\partial y}, \tag{6.2.8}$$

$$\frac{w^{n+1} - w^{n+\sigma}}{\tau} = -\frac{1}{\rho}\frac{\partial P}{\partial z}. \tag{6.2.9}$$

The boundary condition is:

$$\frac{\partial}{\partial n}P(x,y,z,t) = 0. \tag{6.3.6}$$

The calculated cells are cubes that can be completely filled, partially filled, or empty. The cell centers are divided by the distances,, and in x, y, and z, respectively

The cell centers are divided by the distances $h_x/2$, $h_y/2$ и $h_z/2$ по respectively x, y, z. Denote the degree of filling cell (i, j, k) by $o_{i,j,k}$ (polyhedra at the cell centers, Figure 6.3.1).

The velocity and pressure fields are computed at the vertices of the given cell (Figure 6.3.1).

$$(i,j,k),\ (i-1,j,k),\ (i,j-1,k),\ (i-1,j-1,k),$$
$$(i,j,k-1),\ (i,j-1,k-1),\ (i-1,j,k-1),\ (i-1,j-1,k-1).$$

The degree of filling of a cell is determined by the pressure of the water column in this cell. If the mean pressure at the cell nodes is higher than the water column pressure in the cell, then the cell is considered completely filled $o_{i,j,k} = 1$. In the general case, the filling of a cell can be calculated from the following expression:

$$o_{i,j,k} = \frac{\left(\begin{array}{l} P_{i,j,k}H(P_{i,j,k}) + P_{i-1,j,k}H(P_{i-1,j,k}) + P_{i,j-1,k}H(P_{i,j-1,k}) + \\ +P_{i-1,j-1,k}H(P_{i-1,j-1,k}) + P_{i,j,k-1}H(P_{i,j,k-1}) + P_{i-1,j,k-1}H(P_{i-1,j,k-1}) + \\ +P_{i,j-1,k-1}H(P_{i,j-1,k-1}) + P_{i-1,j-1,k-1}H(P_{i-1,j-1,k-1}) \end{array}\right)}{4\rho g h_x}, \tag{6.3.7}$$

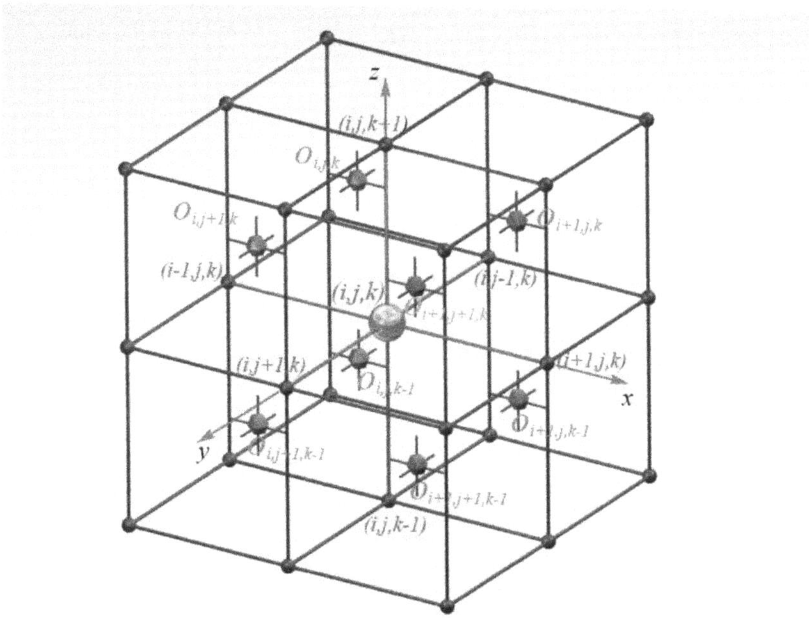

Figure 6.3.1 Geometry of a three-dimensional cell

where $H(x)=\begin{cases}1, & x\geq 0,\\ 0, & x<0\end{cases}$ - Heaviside function.

In the neighborhood of the node (i, j, k) node there are cells

$$(i,j,k),\ (i+1,j,k),\ (i,j+1,k),\ (i+1,j+1,k),$$

$$(i,j,k+1),\ (i,j+1,k+1),\ (i+1,j,k+1),\ (i+1,j+1,k+1)$$

as shown in Figure 6.3.1.

To describe the degree of filling of a domain near a cell, we introduce coefficients: k_0, k_1, k_2, k_3, k_4, k_5, k_6, characterizing the degree of filling of the domains Θ_m:

$$k_0-\Theta_0: x\in(x_{i-1},x_{i+1}),\ \ y\in(y_{j-1},y_{j+1}),\ \ z\in(z_{k-1},z_{k+1});$$

$$k_1-\Theta_1: x\in(x_i,x_{i+1}),\ \ y\in(y_{j-1},y_{j+1}),\ \ z\in(z_{k-1},z_{k+1});$$

$$k_2-\Theta_2: x\in(x_i,x_{i+1}),\ \ y\in(y_j,y_{j+1}),\ \ z\in(z_{k-1},z_{k+1});$$

$$k_3-\Theta_3: x\in(x_{i-1},x_{i+1}),\ \ y\in(y_j,y_{j+1}),\ \ z\in(z_{k-1},z_{k+1}); \quad (6.3.8)$$

$$k_4 - \Theta_4 : x \in (x_{i-1}, x_{i+1}),\ \ y \in (y_{j-1}, y_j),\ \ \ z \in (z_{k-1}, z_{k+1});$$
$$k_5 - \Theta_5 : x \in (x_{i-1}, x_{i+1}),\ \ y \in (y_{j-1}, y_{j+1}),\ \ z \in (z_k, z_{k+1});$$
$$k_6 - \Theta_6 : x \in (x_{i-1}, x_{i+1}),\ \ y \in (y_{j-1}, y_{j+1}),\ \ z \in (z_{k-1}, z_k).$$

Denote by D_m the filled parts of the domains Θ_m, where $m = \overline{0..6}$. Then the coefficients k_m can be calculated by the following expressions:

$$(k_m)_{i,j,k} = \frac{V_{D_m}}{V_{\Theta_m}};$$

$$(k_0)_{i,j,k} = \frac{o_{i,j,k} + o_{i+1,j,k} + o_{i,j+1,k+1} + o_{i+1,j+1,k} + o_{i+1,j,k+1} + o_{i,j+1,k+1} + o_{i+1,j+1,k+1}}{8};$$

$$(k_1)_{i,j,k} = \frac{o_{i+1,j,k} + o_{i+1,j+1,k} + o_{i+1,j,k+1} + o_{i+1,j+1,k+1}}{4};$$

$$(k_2)_{i,j,k} = \frac{o_{i,j,k} + o_{i,j+1,k} + o_{i,j,k+1} + o_{i,j+1,k+1}}{4};$$

$$(k_3)_{i,j,k} = \frac{o_{i,j+1,k} + o_{i+1,j+1,k} + o_{i,j+1,k+1} + o_{i+1,j+1,k+1}}{4}, \tag{6.3.9}$$

$$(k_4)_{i,j,k} = \frac{o_{i,j,k} + o_{i+1,j,k} + o_{i,j,k+1} + o_{i+1,j,k+1}}{4};$$

$$(k_5)_{i,j,k} = \frac{o_{i,j,k+1} + o_{i+1,j,k+1} + o_{i,j+1,k+1} + o_{i+1,j+1,k+1}}{4};$$

$$(k_6)_{i,j,k} = \frac{o_{i,j,k} + o_{i+1,j,k} + o_{i,j+1,k} + o_{i+1,j+1,k}}{4}.$$

We integrate over the domain D_0 (the filled volume of the domain Θ_0) equation (6.2.4):

$$\iiint\limits_{D_0} \frac{u^{n+\sigma} - u^n}{\tau} dxdydz + \iiint\limits_{D_0} u \frac{\partial u}{\partial x} dxdydz + \iiint\limits_{D_0} v \frac{\partial u}{\partial y} dxdydz + \iiint\limits_{D_0} w \frac{\partial u}{\partial z} dxdydz \tag{6.3.10}$$

$$= \iiint\limits_{D_0} \mu \frac{\partial^2 u}{\partial x^2} dxdydz + \iiint\limits_{D_0} \mu \frac{\partial^2 u}{\partial y^2} dxdydz + \iiint\limits_{D_0} \eta \frac{\partial^2 u}{\partial z^2} dxdydz .$$

We compute each integral on the left side separately:

$$\iiint\limits_{D_0} \frac{u^{n+\sigma} - u^n}{\tau} dxdydz = (k_0)_{i,j,k} \iiint\limits_{D_0} \frac{u^{n+\sigma} - u^n}{\tau} dxdydz$$

$$= (k_0)_{i,j,k} \frac{u^{n+\sigma} - u^n}{\tau} h_x h_y h_z \tag{6.3.11}$$

For the second integral in the left side, we obtain:

$$\iiint\limits_{D_0} u \frac{\partial u}{\partial x} dxdydz$$

$$= \frac{(k_1)_{i,j,k} u_{i+1/2,j,k} h_y h_z (u_{i+1,j,k} - u_{i,j,k}) + (k_2)_{i,j,k} u_{i-1/2,j,k} h_y h_z (u_{i,j,k} - u_{i-1,j,k})}{2}. \tag{6.3.12}$$

Similarly for the third and fourth integrals on the left side we obtain:

$$\iiint\limits_{D_0} v \frac{\partial u}{\partial y} dxdydz$$

$$= \frac{(k_3)_{i,j,k} v_{i,j+1/2,k} h_x h_z (u_{i,j+1,k} - u_{i,j,k}) + (k_4)_{i,j,k} v_{i,j-1/2,k} h_x h_z (u_{i,j,k} - u_{i,j-1,k})}{2}. \tag{6.3.13}$$

$$\iiint\limits_{D_0} w \frac{\partial u}{\partial z} dxdydz$$

$$= \frac{(k_5)_{i,j,k} w_{i,j,k+1/2} h_y h_x (u_{i,j,k+1} - u_{i,j,k}) + (k_6)_{i,j,k} w_{i,j,k-1/2} h_y h_x (u_{i,j,k} - u_{i,j,k-1})}{2}. \tag{6.3.14}$$

Let us write the first integral in the right side of the expression (6.3.10)

$$\iiint\limits_{D_0} \mu \frac{\partial^2 u}{\partial x^2} dxdydz = \iiint\limits_{D_1} \mu \frac{\partial^2 u}{\partial x^2} dxdydz + \iiint\limits_{D_2} \mu \frac{\partial^2 u}{\partial x^2} dxdydz . \tag{6.3.15}$$

In this expression, we assume that $V_{D_1} > V_{D_2}$, we select from the region D_1 a fragment adjacent $D_{1,2}$ to the region D_2, at that $V_{D_1} = V_{D_{1,2}}$

$$\iiint\limits_{D_0} \mu \frac{\partial^2 u}{\partial x^2} dxdydz$$

$$= \iiint\limits_{D_1/D_{1,2}} \left(\mu \frac{\partial^2 u}{\partial x^2} \right) dxdydz + \iiint\limits_{D_{1,2}} \left(\mu \frac{\partial^2 u}{\partial x^2} \right) dxdydz + \iiint\limits_{D_2} \left(\mu \frac{\partial^2 u}{\partial x^2} \right) dxdydz \,;$$

$$\iiint\limits_{D_0} \mu \frac{\partial^2 u}{\partial x^2} dxdydz = \iiint\limits_{D_1/D_{1,2}} \left(\mu \frac{\partial^2 u}{\partial x^2} \right) dxdydz$$
$$+(k_2)_{i,j,k} \left(\mu_{i+1/2,j,k} \frac{u_{i+1,j,k} - u_{i,j,k}}{h_x} - \mu_{i-1/2,j,k} \frac{u_{i,j,k} - u_{i-1,j,k}}{h_x} \right), \tag{6.3.16}$$

We calculate the integral over the domain $D_1/D_{1,2}$:

$$\iiint\limits_{D_1/D_{1,2}} \left(\mu \frac{\partial^2 u}{\partial x^2} \right) dxdydz = \int\limits_{z^{(1)}}^{z^{(2)}} dz \int\limits_{y^{(1)}}^{y^{(2)}} dz \int\limits_{x_i}^{x_{i+1/2}} \left(\mu \frac{\partial^2 u}{\partial x^2} \right) dx$$
$$= \left(\mu_{i+1/2,j,k} \frac{u_{i+1,j,k} - u_{i,j,k}}{h_x} - \mu_{i,j,k} (\alpha_{u,x} u_{i,j,k} + \beta_{u,x}) \right)$$
$$\left((k_1)_{i,j,k} - (k_2)_{i,j,k} \right) h_y h_z, \tag{6.3.17}$$

We substitute the expression (6.3.17) into (6.3.16):

$$\iiint\limits_{D_0} \mu \frac{\partial^2 u}{\partial x^2} dxdydz = \left((k_1)_{i,j,k} \mu_{i+1/2,j,k} \frac{u_{i+1,j,k} - u_{i,j,k}}{h_x} - (k_2)_{i,j,k} \mu_{i-1/2,j,k} \right.$$
$$\left. \frac{u_{i,j,k} - u_{i-1,j,k}}{h_x} - \left((k_1)_{i,j,k} - (k_2)_{i,j,k} \right) \mu_{i,j,k} (\alpha_{u,x} u_{i,j,k} + \beta_{u,x}) \right) h_y h_z. \tag{6.3.18}$$

In the case where the $V_{D_1} < V_{D_2}$ result is similar.

For the second integral on the right side of expression (6.3.10), similarly to (6.3.18), we obtain:

$$\iiint_{D_0} \mu \frac{\partial^2 u}{\partial y^2} dxdydz = \left((k_3)_{i,j,k} \mu_{i,j+1/2,k} \frac{u_{i,j+1,k} - u_{i,j,k}}{h_y} - (k_4)_{i,j,k} \mu_{i,j-1/2,k} \right.$$

$$\left. \frac{u_{i,j,k} - u_{i,j-1,k}}{h_y} - \left((k_3)_{i,j,k} - (k_4)_{i,j,k} \right) \mu_{i,j,k} (\alpha_{u,y} u_{i,j,k} + \beta_{u,y}) \right) h_x h_z .$$

(6.3.19)

For the third integral on the right side of expression (6.3.10), similarly to (6.3.18), we obtain:

$$\iiint_{D_0} \eta \frac{\partial^2 u}{\partial z^2} dxdydz = \left((k_5)_{i,j,k} \mu_{i,j,k+1/2} \frac{u_{i,j,k+1} - u_{i,j,k}}{h_z} - (k_6)_{i,j,k} \mu_{i,j,k-1/2} \right.$$

$$\left. \frac{u_{i,j,k} - u_{i,j,k-1}}{h_z} - \left((k_5)_{i,j,k} - (k_6)_{i,j,k} \right) \mu_{i,j,k} (\alpha_{u,z} u_{i,j,k} + \beta_{u,z}) \right) h_y h_x$$

(6.3.20)

We substitute the obtained solutions (6.3.11), (6.3.12), (6.3.13), (6.3.14) and (6.3.18), (6.3.19), (6.3.20) into the original equation (6.3. 10) (taking into account the division by the cell volume $h_x h_y h_z$):

$$(k_0)_{i,j,k} \frac{u^{n+\sigma} - u^n}{\tau} + \tag{6.3.21}$$

$$+ \frac{(k_1)_{i,j,k} u_{i+1/2,j,k} (u_{i+1,j,k} - u_{i,j,k}) + (k_2)_{i,j,k} u_{i-1/2,j,k} (u_{i,j,k} - u_{i-1,j,k})}{2h_x} +$$

$$+ \frac{(k_3)_{i,j,k} v_{i,j+1/2,kz} (u_{i,j+1,k} - u_{i,j,k}) + (k_4)_{i,j,k} v_{i,j-1/2,k} (u_{i,j,k} - u_{i,j-1,k})}{2h_y} +$$

$$+ \frac{(k_5)_{i,j,k} w_{i,j,k+1/2} (u_{i,j,k+1} - u_{i,j,k}) + (k_6)_{i,j,k} w_{i,j,k-1/2} (u_{i,j,k} - u_{i,j,k-1})}{2h_z} =$$

$$=\left((k_1)_{i,j,k}\mu_{i+1/2,j,k}\frac{u_{i+1,j,k}-u_{i,j,k}}{h_x^2}-(k_2)_{i,j,k}\mu_{i-1/2,j,k}\frac{u_{i,j,k}-u_{i-1,j,k}}{h_x^2}-\right.$$

$$\left.-\left((k_1)_{i,j,k}-(k_2)_{i,j,k}\right)\mu_{i,j,k}\frac{(\alpha_{u,x}u_{i,j,k}+\beta_{u,x})}{h_x}\right)+\left((k_3)_{i,j,k}\mu_{i,j+1/2,k}\right.$$

$$\frac{u_{i,j+1,k}-u_{i,j,k}}{h_y^2}-(k_4)_{i,j,k}\mu_{i,j-1/2,k}\frac{u_{i,j,k}-u_{i,j-1,k}}{h_y^2}-\left((k_3)_{i,j,k}-(k_4)_{i,j,k}\right)$$

$$\left.\mu_{i,j,k}\frac{(\alpha_{u,y}u_{i,j,k}+\beta_{u,y})}{h_y}\right)+\left((k_5)_{i,j,k}\eta_{i,j,k+1/2}\frac{u_{i,j,k+1}-u_{i,j,k}}{h_z^2}-(k_6)_{i,j,k}\right.$$

$$\left.\eta_{i,j,k-1/2}\frac{u_{i,j,k}-u_{i,j,k-1}}{h_z^2}-\left((k_5)_{i,j,k}-(k_6)_{i,j,k}\right)\eta_{i,j,k}\frac{(\alpha_{u,z}u_{i,j,k}+\beta_{u,z})}{h_z}\right).$$

Similarly, it is possible to obtain an approximation of the second equation (6.2.5) of the first subtask:

$$(k_0)_{i,j,k}\frac{v^{n+\sigma}-v^n}{\tau}+ \tag{6.3.22}$$

$$+\frac{(k_1)_{i,j,k}u_{i+1/2,j,k}(v_{i+1,j,k}-v_{i,j,k})+(k_2)_{i,j,k}u_{i-1/2,j,k}(v_{i,j,k}-v_{i-1,j,k})}{2h_x}+$$

$$+\frac{(k_3)_{i,j,k}v_{i,j+1/2,k}(v_{i,j+1,k}-v_{i,j,k})+(k_4)_{i,j,k}v_{i,j-1/2,k}(v_{i,j,k}-v_{i,j-1,k})}{2h_y}+$$

$$+\frac{(k_5)_{i,j,k}w_{i,j,k+1/2}(v_{i,j,k+1}-v_{i,j,k})+(k_6)_{i,j,k}w_{i,j,k-1/2}(v_{i,j,k}-v_{i,j,k-1})}{2h_z}=$$

$$=\left((k_1)_{i,j,k}\mu_{i+1/2,j,k}\frac{v_{i+1,j,k}-v_{i,j,k}}{h_x^2}-(k_2)_{i,j,k}\mu_{i-1/2,j,k}\frac{v_{i,j,k}-v_{i-1,j,k}}{h_x^2}\right.$$

$$\left.-\left((k_1)_{i,j,k}-(k_2)_{i,j,k}\right)\mu_{i,j,k}\frac{(\alpha_{v,x}v_{i,j,k}+\beta_{v,x})}{h_x}\right)+\left((k_3)_{i,j,k}\mu_{i,j+1/2,k}\right.$$

$$\frac{v_{i,j+1,k}-v_{i,j,k}}{h_y^2}-+\left((k_3)_{i,j,k}\mu_{i,j+1/2,k}\frac{v_{i,j+1,k}-v_{i,j,k}}{h_y^2}-(k_4)_{i,j,k}\mu_{i,j-1/2,k}\right.$$

$$\left.\frac{v_{i,j,k}-v_{i,j-1,k}}{h_y^2}-\left((k_3)_{i,j,k}-(k_4)_{i,j,k}\right)\mu_{i,j,k}\frac{(\alpha_{v,y}v_{i,j,k}+\beta_{v,y})}{h_y}\right)+$$

$$+\left((k_5)_{i,j,k}\eta_{i,j,k+1/2}\frac{v_{i,j,k+1}-v_{i,j,k}}{h_z^2}-(k_6)_{i,j,k}\eta_{i,j,k-1/2}\frac{v_{i,j,k}-v_{i,j,k-1}}{h_z^2}\right.$$

$$\left.-\left((k_5)_{i,j,k}-(k_6)_{i,j,k}\right)\eta_{i,j,k}\frac{(\alpha_{v,z}v_{i,j,k}+\beta_{v,z})}{h_z}\right).$$

Similarly, for the third equation (6.2.6) of the first subtask:

$$(k_0)_{i,j,k}\frac{w^{n+\sigma}-w^n}{\tau}+ \tag{6.3.23}$$

$$+\frac{(k_1)_{i,j,k}u_{i+1/2,j,k}(w_{i+1,j,k}-w_{i,j,k})+(k_2)_{i,j,k}u_{i-1/2,j,k}(w_{i,j,k}-w_{i-1,j,k})}{2h_x}+$$

$$+\frac{(k_3)_{i,j,k}v_{i,j+1/2,k}(w_{i,j+1,k}-w_{i,j,k})+(k_4)_{i,j,k}v_{i,j-1/2,k}(w_{i,j,k}-w_{i,j-1,k})}{2h_y}+$$

$$+\frac{(k_5)_{i,j,k}w_{i,j,k+1/2}(w_{i,j,k+1}-w_{i,j,k})+(k_6)_{i,j,k}w_{i,j,k-1/2}(w_{i,j,k}-w_{i,j,k-1})}{2h_z}=$$

$$=\left((k_1)_{i,j,k}\eta_{i+1/2,j,k}\frac{w_{i+1,j,k}-w_{i,j,k}}{h_x^2}-(k_2)_{i,j,k}\eta_{i-1/2,j,k}\frac{w_{i,j,k}-w_{i-1,j,k}}{h_x^2}\right.$$

$$\left.-\left((k_1)_{i,j,k}-(k_2)_{i,j,k}\right)\eta_{i,j,k}\frac{(\alpha_{w,x}w_{i,j,k}+\beta_{w,x})}{h_x}\right)+$$

$$+\left((k_3)_{i,j,k}\eta_{i,j+1/2,k}\frac{w_{i,j+1,k}-w_{i,j,k}}{h_y^2}-(k_4)_{i,j,k}w_{i,j-1/2,k}\frac{w_{i,j,k}-w_{i,j-1,k}}{h_y^2}\right.$$

$$-\left((k_3)_{i,j,k}-(k_4)_{i,j,k}\right)\eta_{i,j,k}\frac{(\alpha_{w,y}w_{i,j,k}+\beta_{w,y})}{h_y}\Bigg)+\Big((k_5)_{i,j,k}\eta_{i,j,k+1/2}$$

$$\frac{w_{i,j,k+1}-w_{i,j,k}}{h_z^2}-(k_6)_{i,j,k}\eta_{i,j,k-1/2}\frac{w_{i,j,k}-w_{i,j,k-1}}{h_z^2}-$$

$$\left((k_5)_{i,j,k}-(k_6)_{i,j,k}\right)\eta_{i,j,k}\frac{(\alpha_{w,z}w_{i,j,k}+\beta_{w,z})}{h_z}\Bigg)+(k_0)_{i,j,k}g$$

In the case of internal nodes and boundary nodes located on the surface of the reservoir and the lateral boundary:

$$\begin{aligned}\alpha_{u,x}=0,\ \alpha_{v,y}=0,\ \alpha_{w,z}=0,\\ \beta_{u,x}=0,\ \beta_{v,y}=0,\ \beta_{w,z}=0.\end{aligned}\tag{6.3.24}$$

For these nodes on the right side of the equations, the last term is converted to zero. For the boundary nodes located at the bottom of the region under consideration and describing the motion of the fluid, equation (6.3.21) takes the form:

$$(k_0)_{i,j,k}\frac{u^{n+\sigma}-u^n}{\tau}+\tag{6.3.25}$$

$$+\frac{(k_1)_{i,j,k}u_{i+1/2,j,k}(u_{i+1,j,k}-u_{i,j,k})+(k_2)_{i,j,k}u_{i-1/2,j,k}(u_{i,j,k}-u_{i-1,j,k})}{2h_x}+$$

$$+\frac{(k_3)_{i,j,k}v_{i,j+1/2,kz}(u_{i,j+1,k}-u_{i,j,k})+(k_4)_{i,j,k}v_{i,j-1/2,k}(u_{i,j,k}-u_{i,j-1,k})}{2h_y}+$$

$$+\frac{(k_5)_{i,j,k}w_{i,j,k+1/2}(u_{i,j,k+1}-u_{i,j,k})+(k_6)_{i,j,k}w_{i,j,k-1/2}(u_{i,j,k}-u_{i,j,k-1})}{2h_z}=$$

$$=\left((k_1)_{i,l,k}\mu_{i+1/2,j,k}\frac{u_{i+1,j,k}-u_{i,j,k}}{h_x^2}-(k_2)_{i,l,k}\mu_{i-1/2,j,k}\frac{u_{i,j,k}-u_{i-1,j,k}}{h_x^2}-\right)+$$

$$+\Bigg((k_3)_{i,l,k}\mu_{i,j+1/2,k}\frac{u_{i,j+1,k}-u_{i,j,k}}{h_y^2}-(k_4)_{i,l,k}\mu_{i,j-1/2,k}\frac{u_{i,j,k}-u_{i,j-1,k}}{h_y^2}$$

$$-\left((k_3)_{i,l,k}-(k_4)_{i,l,k}\right)\frac{\tau_x}{\rho h_y}\Bigg)+\Bigg((k_5)_{i,l,k}\eta_{i,j,k+1/2}\frac{u_{i,j,k+1}-u_{i,j,k}}{h_z^2}$$

$$-(k_6)_{i,l,k}\eta_{i,j,k-1/2}\frac{u_{i,j,k}-u_{i,j,k-1}}{h_z^2}\Bigg).$$

For boundary nodes located at the bottom of the region, equation (6.3.22) takes the form:

$$(k_0)_{i,j,k}\frac{v^{n+\sigma}-v^n}{\tau}+ \tag{6.3.26}$$

$$+\frac{(k_1)_{i,j,k}u_{i+1/2,j,k}(v_{i+1,j,k}-v_{i,j,k})+(k_2)_{i,j,k}u_{i-1/2,j,k}(v_{i,j,k}-v_{i-1,j,k})}{2h_x}+$$

$$+\frac{(k_3)_{i,j,k}v_{i,j+1/2,k}(v_{i,j+1,k}-v_{i,j,k})+(k_4)_{i,j,k}v_{i,j-1/2,k}(v_{i,j,k}-v_{i,j-1,k})}{2h_y}+$$

$$+\frac{(k_5)_{i,j,k}w_{i,j,k+1/2}(v_{i,j,k+1}-v_{i,j,k})+(k_6)_{i,j,k}w_{i,j,k-1/2}(v_{i,j,k}-v_{i,j,k-1})}{2h_z}=$$

$$=\Bigg((k_1)_{i,l,k}\mu_{i+1/2,j,k}\frac{v_{i+1,j,k}-v_{i,j,k}}{h_x^2}-(k_2)_{i,l,k}\mu_{i-1/2,j,k}\frac{v_{i,j,k}-v_{i-1,j,k}}{h_x^2}-$$

$$\left((k_1)_{i,l,k}-(k_2)_{i,l,k}\right)\frac{\tau_y}{\rho h_x}\Bigg)+\Bigg((k_3)_{i,l,k}\mu_{i,j+1/2,k}\frac{v_{i,j+1,k}-v_{i,j,k}}{h_y^2}-$$

$$(k_4)_{i,l,k}\mu_{i,j-1/2,k}\frac{v_{i,j,k}-v_{i,j-1,k}}{h_y^2}\Bigg)+\Bigg((k_5)_{i,l,k}\eta_{i,j,k+1/2}\frac{v_{i,j,k+1}-v_{i,j,k}}{h_z^2}$$

$$-(k_6)_{i,l,k}\eta_{i,j,k-1/2}\frac{v_{i,j,k}-v_{i,j,k-1}}{h_z^2}\Bigg).$$

For boundary nodes located at the bottom of the region, equation (6.3.23):

$$(k_0)_{i,j,k}\frac{w^{n+\sigma}-w^n}{\tau}+ \tag{6.3.27}$$

$$+\frac{(k_3)_{i,j,k}v_{i,j+1/2,k}(w_{i,j+1,k}-w_{i,j,k})+(k_4)_{i,j,k}v_{i,j-1/2,k}(w_{i,j,k}-w_{i,j-1,k})}{2h_y}+$$

$$+\frac{(k_5)_{i,j,k}w_{i,j,k+1/2}(w_{i,j,k+1}-w_{i,j,k})+(k_6)_{i,j,k}w_{i,j,k-1/2}(w_{i,j,k}-w_{i,j,k-1})}{2h_z}=$$

$$=\left((k_1)_{i,l,k}\eta_{i+1/2,j,k}\frac{w_{i+1,j,k}-w_{i,j,k}}{h_x^2}-(k_2)_{i,l,k}\eta_{i-1/2,j,k}\frac{w_{i,j,k}-w_{i-1,j,k}}{h_x^2}\right)+$$

$$+\left((k_3)_{i,l,k}\eta_{i,j+1/2,k}\frac{w_{i,j+1,k}-w_{i,j,k}}{h_y^2}-(k_4)_{i,l,k}w_{i,j-1/2,k}\frac{w_{i,j,k}-w_{i,j-1,k}}{h_y^2}\right)+$$

$$+\left((k_5)_{i,l,k}\eta_{i,j,k+1/2}\frac{w_{i,j,k+1}-w_{i,j,k}}{h_z^2}-(k_6)_{i,l,k}\eta_{i,j,k-1/2}\frac{w_{i,j,k}-w_{i,j,k-1}}{h_z^2}\right.$$

$$\left.-\left((k_5)_{i,l,k}-(k_6)_{i,l,k}\right)\frac{\tau_z}{\rho h_z}\right)+(k_0)_{i,j,k}g\,.$$

We now turn to the second subtask, construct a finite-volume model on the basis of the equation for the pressure (6.2.10), taking into account the following boundary conditions:

$$\frac{\partial P(x,y,z,t)}{\partial n}=\alpha_{p,n}P+\beta_{p,n},$$

$$u(x,y,z,t)=u^{(0)},\quad v(x,y,z,t)=v^{(0)},\quad w(x,y,z,t)=w^{(0)}. \qquad (6.3.28)$$

We integrate the equation (6.2.10) over the domain D_0:

$$\iiint\limits_{D_0}\frac{\partial^2 P}{\partial x^2}dxdydz+\iiint\limits_{D_0}\frac{\partial^2 P}{\partial y^2}dxdydz+\iiint\limits_{D_0}\frac{\partial^2 P}{\partial z^2}dxdydz=$$

$$=\frac{\rho}{\tau}\left(\iiint\limits_{D_0}\frac{\partial u^{n+\sigma}}{\partial x}dxdydz+\iiint\limits_{D_0}\frac{\partial v^{n+\sigma}}{\partial y}dxdydz+\iiint\limits_{D_0}\frac{\partial w^{n+\sigma}}{\partial z}dxdydz\right).$$

(6.3.29)

We calculate separately each integral:

$$\iiint\limits_{D_0} \frac{\partial^2 P}{\partial x^2} dxdydz = \iiint\limits_{D_1/D_2} \frac{\partial^2 P}{\partial x^2} dxdydz + \iiint\limits_{D_{1,2}} \frac{\partial^2 P}{\partial x^2} dxdydz + \iiint\limits_{D_2} \frac{\partial^2 P}{\partial x^2} dxdydz$$

$$= \iiint\limits_{D_1/D_2} \frac{\partial^2 P}{\partial x^2} dxdydz + (k_2)_{i,j,k} \iiint\limits_{D_2} \frac{\partial^2 P}{\partial x^2} dxdydz$$

$$= \iiint\limits_{D_1/D_2} \frac{\partial^2 P}{\partial x^2} dxdydz + (k_2)_{i,j,k}$$

$$\left(\frac{P_{i+1,j,k} - P_{i,j,k}}{h_x} - \frac{P_{i,j,k} - P_{i-1,j,k}}{h_x} \right) h_y h_z. \tag{6.3.30}$$

We calculate the integral over the domain $D_1/D_{1,2}$:

$$\iiint\limits_{D_1/D_2} \frac{\partial^2 P}{\partial x^2} dxdydz = \int\limits_{z^{(1)}}^{z^{(2)}} dz \int\limits_{y^{(1)}}^{y^{(2)}} dy \int\limits_{x_i}^{x_{i+1/2}} \frac{\partial^2 P}{\partial x^2} dx = \left(\frac{P_{i+1,j,k} - P_{i,j,k}}{h_x} - \right.$$

$$\left. - \left(\alpha_{p,x} P_{i,j,k} + \beta_{p,x} \right) \right) \left((k_1)_{i,j,k} - (k_2)_{i,j,k} \right) h_y h_z. \tag{6.3.31}$$

We substitute the solutions obtained in the original expression (6.3.30):

$$\iiint\limits_{D_0} \frac{\partial^2 P}{\partial x^2} dxdydz = \left(\frac{P_{i+1,j,k} - P_{i,j,k}}{h_x} - \left(\alpha_{p,x} P_{i,j,k} + \beta_{p,x} \right) \right) \left((k_1)_{i,j,k} - (k_2)_{i,j,k} \right)$$

$$h_y h_z + (k_2)_{i,j,k} \left(\frac{P_{i+1,j,k} - P_{i,j,k}}{h_x} - \frac{P_{i,j,k} - P_{i-1,j,k}}{h_x} \right) h_y h_z. \tag{6.3.32}$$

Similarly, the second integral on the left side of expression (6.3.29):

$$\iiint\limits_{D_0} \frac{\partial^2 P}{\partial y^2} dxdydz = \left(\frac{P_{i,j+1,k} - P_{i,j,k}}{h_y} - \left(\alpha_{p,y} P_{i,j,k} + \beta_{p,y} \right) \right)$$

$$\left((k_3)_{i,j,k} - (k_4)_{i,j,k} \right) h_x h_z + (k_4)_{i,j,k} \left(\frac{P_{i,j+1,k} - P_{i,j,k}}{h_y} - \frac{P_{i,j,k} - P_{i,j-1,k}}{h_y} \right) h_x h_z. \tag{6.3.33}$$

For the third integral on the left side of expression (6.3.29):

$$\iiint\limits_{D_0} \frac{\partial^2 P}{\partial z^2} dxdydz = \left(\frac{P_{i,j,k+1} - P_{i,j,k}}{h_z} - \left(\alpha_{p,z} P_{i,j,k} + \beta_{p,z} \right) \right) \left((k_5)_{i,j,k} - (k_6)_{i,j,k} \right) h_x h_y + (k_6)_{i,j,k} \left(\frac{P_{i,j,k+1} - P_{i,j,k}}{h_z} - \frac{P_{i,j,k} - P_{i,j,k-1}}{h_z} \right) h_x h_y. \tag{6.3.34}$$

We now turn to the calculation of the first integral on the right side of expression (6.3.29):

$$\iiint\limits_{D_0} \frac{\partial u^{n+\sigma}}{\partial x} dxdydz = \iiint\limits_{D_1} \frac{\partial u^{n+\sigma}}{\partial x} dxdydz + \iiint\limits_{D_2} \frac{\partial u^{n+\sigma}}{\partial x} dxdydz = \iiint\limits_{D_1/D_2} \frac{\partial u^{n+\sigma}}{\partial x} dxdydz + (k_2)_{i,j,k} \left(u^{n+\sigma}_{i+1/2,j,k} - u^{n+\sigma}_{i-1/2,j,k} \right) h_y h_z. \tag{6.3.35}$$

Calculate the integral over the domain $D_1/D_{1,2}$:

$$\iiint\limits_{D_1/D_2} \frac{\partial u^{n+\sigma}}{\partial x} dxdydz = \int\limits_{z^{(1)}}^{z^{(2)}} dz \int\limits_{y_i}^{y^{(2)}} dy \int\limits_{x_i}^{x_{i+1/2}} \left(\frac{\partial u^{n+\sigma}}{\partial x} \right) dx = \left((k_1)_{i,j,k} - (k_2)_{i,j,k} \right) \left(u^{n+\sigma}_{i+1/2,j,k} - (u^{(0)})_{i,j,k} \right) h_y h_z. \tag{6.3.36}$$

Substitute the solutions obtained in the original expression (6.3.35):

$$\iiint\limits_{D_0} \frac{\partial u^{n+\sigma}}{\partial x} dxdydz = \left((k_1)_{i,j,k,} u^{n+\sigma}_{i+1/2,j,k} - (k_2)_{i,j,k} u^{n+\sigma}_{i-1/2,j,k} - \left((k_1)_{i,j,k} - (k_2)_{i,j,k} \right) (u^{(0)})_{i,j,k} \right) h_y h_z. \tag{6.3.37}$$

After analogous transformations for the second integral on the right side of expression (6.3.29), we obtain:

$$\iiint_{D_0} \frac{\partial v^{n+\sigma}}{\partial y} dxdydz$$
$$= \left((k_3)_{i,j,k,} v^{n+\sigma}_{i,j+1/2,k} - (k_4)_{i,j,k} v^{n+\sigma}_{i,j-1/2,k} - \left((k_3)_{i,j,k} - (k_4)_{i,j,k}\right)(v^{(0)})_{i,j,k}\right) h_x h_z .$$
(6.3.38)

For the third integral we obtain:

$$\iiint_{D_0} \frac{\partial w^{n+\sigma}}{\partial z} dxdydz$$
$$= \left((k_5)_{i,j,k,} w^{n+\sigma}_{i,j,k+1/2} - (k_6)_{i,j,k} w^{n+\sigma}_{i,j,k-1/2} - \left((k_5)_{i,j,k} - (k_6)_{i,j,k}\right)(w^{(0)})_{i,j,k}\right) h_x h_y .$$
(6.3.39)

Substitute the results of integration into the original expression (6.3.29) (taking into account the division by the cell volume $h_x h_y h_z$):

$$(k_1)_{i,j,k} \frac{P_{i+1,j,k} - P_{i,j,k}}{h_x^2} - (k_2)_{i,j,k} \frac{P_{i,j,k} - P_{i-1,j,k}}{h_x^2} - \frac{\alpha_{p,x} P_{i,j,k} + \beta_{p,x}}{h_x}$$

$$\left((k_1)_{i,j,k} - (k_2)_{i,j,k}\right) + (k_3)_{i,j,k} \frac{P_{i,j+1,k} - P_{i,j,k}}{h_y^2} - (k_4)_{i,j,k} \frac{P_{i,j,k} - P_{i,j-1,k}}{h_y^2}$$

$$- \frac{\alpha_{p,x} P_{i,j,k} + \beta_{p,x}}{h_y} \left((k_3)_{i,j,k} - (k_4)_{i,j,k}\right) + (k_5)_{i,j,k} \frac{P_{i,j,k+1} - P_{i,j,k}}{h_z^2}$$

$$-(k_6)_{i,j,k} \frac{P_{i,j,k} - P_{i,j,k-1}}{h_z^2} - \frac{\alpha_{p,x} P_{i,j,k} + \beta_{p,x}}{h_z} \left((k_5)_{i,j,k} - (k_6)_{i,j,k}\right)$$

$$= \frac{\rho}{\tau} \left(\frac{(k_1)_{i,j,k} u^{n+\sigma}_{i+1/2,j,k} - (k_2)_{i,j,k} u^{n+\sigma}_{i-1/2,j,k} - \left((k_1)_{i,j,k} - (k_2)_{i,j,k}\right)(u^{(0)})_{i,j,k}}{h_x} \right) +$$

$$+ \frac{\rho}{\tau} \left(\frac{(k_3)_{i,j,k} v^{n+\sigma}_{i,j+1/2,k} - (k_4)_{i,j,k} v^{n+\sigma}_{i,j-1/2,k} - \left((k_3)_{i,j,k} - (k_4)_{i,j,k}\right)(v^{(0)})_{i,j,k}}{h_y} \right) +$$

$$+\frac{\rho}{\tau}\left(\frac{(k_5)_{i,j,k}w^{n+\sigma}_{i,j,k+1/2}-(k_6)_{i,j,k}w^{n+\sigma}_{i,j,k-1/2}-\left((k_5)_{i,j,k}-(k_6)_{i,j,k}\right)(w^{(0)})_{i,j,k}}{h_z}\right). \tag{6.3.40}$$

Next, we integrate the boundary condition on the free surface of the liquid (6.3.4) for the vertical component of the particle velocity in the region $D:\left\{\left(x,y,z\right)\in D_0;\ t\in\left(t^n,t^{n+1}\right)\right\}$:

$$\iiint\limits_D\int w(x,y,z,t)dxdydzdt=\iiint\limits_D\int\frac{1}{g\rho}\frac{\partial P}{\partial t}dxdydzdt. \tag{6.3.41}$$

For the integral on the left side of this expression obtain:

$$\iiint\limits_D\int w(x,y,z,t)dxdydzdt=(k_0)_{i,j,k}(w^{(0)})_{i,j,k}h_xh_yh_z\tau. \tag{6.3.42}$$

The integral on the right side takes the form:

$$\begin{aligned}\iiint\limits_D\int\frac{1}{g\rho}\frac{\partial P}{\partial t}dxdydzdt&=\frac{1}{g\rho}\iiint\limits_{D_0}\left(\int\limits_{t^n}^{t^{n+1}}\frac{\partial P}{\partial t}dt\right)dxdydz\\&=(k_0)_{i,j,k}\frac{P^{n+1}_{i,j,k}-P^n_{i,j,k}}{g\rho}h_xh_yh_z.\end{aligned} \tag{6.3.43}$$

Equating the solutions obtained, we finally obtain:

$$(w^{(0)})_{i,j,k}=\frac{P^{n+1}_{i,j,k}-P^n_{i,j,k}}{g\rho\tau}. \tag{6.3.44}$$

In expression (6.3.40), the pressure field will be considered implicitly, in the case of boundary nodes at the bottom and the surface of the region, as well as internal nodes, this expression will take the following form:

$$(k_1)_{i,j,k}\frac{P_{i+1,j,k}-P_{i,j,k}}{h_x^2}-(k_2)_{i,j,k}\frac{P_{i,j,k}-P_{i-1,j,k}}{h_x^2}+$$

$$+(k_3)_{i,j,k}\frac{P_{i,j+1,k}-P_{i,j,k}}{h_y^2}-(k_4)_{i,j,k}\frac{P_{i,j,k}-P_{i,j-1,k}}{h_y^2}+$$

$$+(k_5)_{i,j,k}\frac{P_{i,j,k+1}-P_{i,j,k}}{h_z^2}-(k_6)_{i,j,k}\frac{P_{i,j,k}-P_{i,j,k-1}}{h_z^2}=$$

$$=\frac{\rho}{\tau}\left(\frac{(k_1)_{i,j,k}u^{n+\sigma}_{i+1/2,j,k}-(k_2)_{i,j,k}u^{n+\sigma}_{i-1/2,j,k}}{h_x}\right)+$$

$$+\frac{\rho}{\tau}\left(\frac{(k_3)_{i,j,k}v^{n+\sigma}_{i,j+1/2,k}-(k_4)_{i,j,k}v^{n+\sigma}_{i,j-1/2,k}}{h_y}\right)+$$

$$+\frac{\rho}{\tau}\left(\frac{(k_5)_{i,j,k}w^{n+\sigma}_{i,j,k+1/2}-(k_6)_{i,j,k}w^{n+\sigma}_{i,j,k-1/2}}{h_z}\right)-$$

$$-\frac{\left((k_1)_{i,j,k}-(k_2)_{i,j,k}\right)(P^{n+1}_{i,j,k}-P^{n}_{i,j,k})_{i,j,k}}{gh_x\tau^2}H\left((k_1)_{i,j,k}-(k_2)_{i,j,k}\right). \quad (6.3.44a)$$

wher $H(x)$- Heaviside function. We shall consider nodes for which the surface nodes are $k_1 > k_2$.

The expression (6.3.40) for the boundary nodes at the lateral boundary, taking into account the rise of the liquid level, takes the form:

$$(k_1)_{i,j,k}\frac{P_{i+1,j,k}-P_{i,j,k}}{h_x^2}-(k_2)_{i,j,k}\frac{P_{i,j,k}-P_{i-1,j,k}}{h_x^2}+$$

$$+(k_3)_{i,j,k}\frac{P_{i,j+1,k}-P_{i,j,k}}{h_y^2}-(k_4)_{i,j,k}\frac{P_{i,j,k}-P_{i,j-1,k}}{h_y^2}+$$

$$+(k_5)_{i,j,k}\frac{P_{i,j,k+1}-P_{i,j,k}}{h_z^2}-(k_6)_{i,j,k}\frac{P_{i,j,k}-P_{i,j,k-1}}{h_z^2}-\frac{\alpha(m_2)_{i,j,k}}{h_y}$$

$$\left((k_3)_{i,j,k}-(k_4)_{i,j,k}\right)=\frac{\rho}{\tau}\left(\frac{(k_1)_{i,j,k}u^{n+\sigma}_{i+1/2,j,k}-(k_2)_{i,j,k}u^{n+\sigma}_{i-1/2,j,k}}{h_x}\right)+$$

$$+\frac{\rho}{\tau}\left(\frac{(k_3)_{i,j,k}v^{n+\sigma}_{i,j+1/2,k}-(k_4)_{i,j,k}v^{n+\sigma}_{i,j-1/2,k}}{h_y}\right)+$$

$$+\frac{\rho}{\tau}\left(\frac{(k_5)_{i,j,k}w^{n+\sigma}_{i,j,k+1/2}-(k_6)_{i,j,k}w^{n+\sigma}_{i,j,k-1/2}}{h_z}\right)-$$

$$-\frac{\left((k_1)_{i,j,k}-(k_2)_{i,j,k}\right)(P^{n+1}_{i,j,k}-P^{n}_{i,j,k})_{i,j,k}}{gh_x\tau^2}H\left((k_1)_{i,j,k}-(k_2)_{i,j,k}\right). \quad (6.3.45)$$

Let us proceed further to the third subtask, integrate equation (6.2.7) over the domain D_0:

$$\iiint\limits_{D_0}\frac{u^{n+1}-u^{n+\sigma}}{\tau}dxdydz=-\frac{1}{\rho}\iiint\limits_{D_0}\frac{\partial P}{\partial x}dxdydz. \quad (6.3.46)$$

Compute the integral on the left side of expression (6.3.46):

$$\iiint\limits_{D_0}\frac{u^{n+1}-u^{n+\sigma}}{\tau}dxdydz=(k_0)_{i,j,k}\frac{u^{n+1}_{i,j,k}-u^{n+\sigma}_{i,j,k}}{\tau}h_xh_yh_z. \quad (6.3.47)$$

Consider the second integral of expression (6.3.46):

$$\iiint\limits_{D_0}\frac{\partial P}{\partial x}dxdydz=\iiint\limits_{D_1/D_2}\frac{\partial P}{\partial x}dxdydz+(k_2)_{i,j,k}\iiint\limits_{\Theta_0}\frac{\partial P}{\partial x}dxdydz=$$

$$=\int\limits_{z^{(1)}}^{z^{(2)}}dz\int\limits_{y^{(1)}}^{y^{(2)}}dy\int\limits_{x_i}^{x_{i+1/2}}\frac{\partial P}{\partial x}dx+(k_2)_{i.j}\left(P_{i+1/2,j,k}-P_{i-1/2,j,k}\right)h_yh_z=$$

$$=\left[\left((k_1)_{i.j,k}-(k_2)_{i,j,k}\right)\left(P_{i+1/2,j,k}-P_{i,j,k}\right)+(k_2)_{i,j,k}\left(P_{i+1/2,j,k}-P_{i-1/2,j,k}\right)\right]h_yh_z=$$

$$= \left[(k_1)_{i,j,k} \frac{\left(P_{i+1,j,k} - P_{i,j,k}\right)}{2} + (k_2)_{i,j,k} \frac{\left(P_{i,j,k} - P_{i-1,j,k}\right)}{2} \right] h_y h_z. \quad (6.3.48)$$

We substitute the obtained solutions (6.3.47) and (6.3.48) into the original expression (6.3.46) (taking into account the division by the cell volume $h_x h_y h_z$), the pressure field will be on the time layer $n + 1$:

$$(k_0)_{i,j,k} \frac{u_{i,j,k}^{n+1} - u_{i,j,k}^{n+\sigma}}{\tau} = -\left((k_1)_{i,j,k} \frac{P_{i+1,j,k}^{n+1} - P_{i,j,k}^{n+1}}{2h_x \rho} + (k_2)_{i,j,k} \frac{P_{i,j,k}^{n+1} - P_{i-1,j,k}^{n+1}}{2h_x \rho} \right). \quad (6.3.49)$$

Similarly, one can obtain discrete analogs for the remaining components of the velocity of equations (6.2.8), (6.2.9):

$$(k_0)_{i,j,k} \frac{v_{i,j,k}^{n+1} - v_{i,j,k}^{n+\sigma}}{\tau} = -\left((k_3)_{i,j,k} \frac{P_{i,j+1,k}^{n+1} - P_{i,j,k}^{n+1}}{2h_y \rho} + (k_4)_{i,j,k} \frac{P_{i,j,k}^{n+1} - P_{i,j-1,k}^{n+1}}{2h_y \rho} \right), \quad (6.3.50)$$

$$(k_0)_{i,j,k} \frac{w_{i,j,k}^{n+1} - w_{i,j,k}^{n+\sigma}}{\tau} = -\left((k_5)_{i,j,k} \frac{P_{i,j,k+1}^{n+1} - P_{i,j,k}^{n+1}}{2h_z \rho} + (k_6)_{i,j,k} \frac{P_{i,j,k}^{n+1} - P_{i,j,k-1}^{n+1}}{2h_z \rho} \right). \quad (6.3.51)$$

We introduce the «masks» of the boundary conditions: m_1, m_2. Parameter $m_1 = 1$, if the node (i, j, k) belongs to the set of boundary nodes located in the bottom area, otherwise $m_1 = 0$. Parameter $m_2 = 1$, if the node (i, j, k) belongs to a lateral boundary, otherwise it takes a value $m_2 = 0$.

Taking into account the introduced notations, the discrete finite-volume model will be represented by the following equations:

- grid equation for calculating the component of the velocity vector $u_{i,j,k}$:

$$(k_0)_{i,j,k} \frac{u^{n+\sigma} - u^n}{\tau} + \quad (6.3.52)$$

$$+(k_1)_{i,j,k} u^n_{i+1/2,j,k} \frac{(u^{n+\sigma/2}_{i+1,j,k} - u^{n+\sigma/2}_{i,j,k})}{2h_x} + (k_2)_{i,j,k} u^n_{i-1/2,j,k} \frac{(u^{n+\sigma/2}_{i,j,k} - u^{n+\sigma/2}_{i-1,j,k})}{2h_x} +$$

$$+(k_3)_{i,j,k} v^n_{i,j+1/2,k} \frac{(u^{n+\sigma/2}_{i,j+1,k} - u^{n+\sigma/2}_{i,j,k})}{2h_y} + (k_4)_{i,j,k} v^n_{i,j-1/2,k} \frac{(u^{n+\sigma/2}_{i,j,k} - u^{n+\sigma/2}_{i,j-1,k})}{2h_y} +$$

$$+(k_5)_{i,j,k} w^n_{i,j,k+1/2} \frac{(u^{n+\sigma/2}_{i,j,k+1} - u^{n+\sigma/2}_{i,j,k})}{2h_z} + (k_6)_{i,j,k} w^n_{i,j,k-1/2} \frac{(u^{n+\sigma/2}_{i,j,k} - u^{n+\sigma/2}_{i,j,k-1})}{2h_z} =$$

$$= \left((k_1)_{i,j,k} \mu_{i+1/2,j,k} \frac{u^{n+\sigma/2}_{i+1,j,k} - u^{n+\sigma/2}_{i,j,k}}{h_x^2} - (k_2)_{i,j,k} \mu_{i-1/2,j,k} \frac{u^{n+\sigma/2}_{i,j,k} - u^{n+\sigma/2}_{i-1,j,k}}{h_x^2} \right) +$$

$$+ \left((k_3)_{i,j,k} \mu_{i,j+1/2,k} \frac{u^{n+\sigma/2}_{i,j+1,k} - u^{n+\sigma/2}_{i,j,k}}{h_y^2} - (k_4)_{i,j,k} \mu_{i,j-1/2,k} \frac{u^{n+\sigma/2}_{i,j,k} - u^{n+\sigma/2}_{i,j-1,k}}{h_y^2} \right) +$$

$$+ \left((k_5)_{i,j,k} \eta_{i,j,k+1/2} \frac{u^{n+\sigma/2}_{i,j,k+1} - u^{n+\sigma/2}_{i,j,k}}{h_z^2} - (k_6)_{i,j,k} \eta_{i,j,k-1/2} \frac{u^{n+\sigma/2}_{i,j,k} - u^{n+\sigma/2}_{i,j,k-1}}{h_z^2} \right) +$$

$$+\left((k_3)_{i,j,k} - (k_4)_{i,j,k} \right) \frac{\tau_x}{\rho h_y} (m_1)_{i,j,k};$$

- grid equation for calculating the component of the velocity vector $v_{i,j,k}$:

$$(k_0)_{i,j,k} \frac{v^{n+\sigma} - v^n}{\tau} + \tag{6.3.53}$$

$$+(k_1)_{i,j,k} u^n_{i+1/2,j,k} \frac{(v^{n+\sigma/2}_{i+1,j,k} - v^{n+\sigma/2}_{i,j,k})}{2h_x} + (k_2)_{i,j,k} u^n_{i-1/2,j,k} \frac{(v^{n+\sigma/2}_{i,j,k} - v^{n+\sigma/2}_{i-1,j,k})}{2h_x} +$$

$$+(k_3)_{i,j,k} v^n_{i,j+1/2,k} \frac{(v^{n+\sigma/2}_{i,j+1,k} - v^{n+\sigma/2}_{i,j,k})}{2h_y} + (k_4)_{i,j,k} v^n_{i,j-1/2,k} \frac{(v^{n+\sigma/2}_{i,j,k} - v^{n+\sigma/2}_{i,j-1,k})}{2h_y} +$$

$$+(k_5)_{i,j,k} w^n_{i,j,k+1/2} \frac{(v^{n+\sigma/2}_{i,j,k+1} - v^{n+\sigma/2}_{i,j,k})}{2h_z} + (k_6)_{i,j,k} w^n_{i,j,k-1/2} \frac{(v^{n+\sigma/2}_{i,j,k} - v^{n+\sigma/2}_{i,j,k-1})}{2h_z} =$$

$$= \left((k_1)_{i,j,k} \mu_{i+1/2,j,k} \frac{v^{n+\sigma/2}_{i+1,j,k} - v^{n+\sigma/2}_{i,j,k}}{h_x^2} - (k_2)_{i,j,k} \mu_{i-1/2,j,k} \frac{v^{n+\sigma/2}_{i,j,k} - v^{n+\sigma/2}_{i-1,j,k}}{h_x^2} \right) +$$

$$+ \left((k_3)_{i,j,k} \mu_{i,j+1/2,k} \frac{v^{n+\sigma/2}_{i,j+1,k} - v^{n+\sigma/2}_{i,j,k}}{h_y^2} - (k_4)_{i,j,k} \mu_{i,j-1/2,k} \frac{v^{n+\sigma/2}_{i,j,k} - v^{n+\sigma/2}_{i,j-1,k}}{h_y^2} \right) +$$

$$+ \left((k_5)_{i,j,k} \eta_{i,j,k+1/2} \frac{v^{n+\sigma/2}_{i,j,k+1} - v^{n+\sigma/2}_{i,j,k}}{h_z^2} - (k_6)_{i,j,k} \eta_{i,j,k-1/2} \frac{v^{n+\sigma/2}_{i,j,k} - v^{n+\sigma/2}_{i,j,k-1}}{h_z^2} \right) +$$

$$+ \left((k_1)_{i,j,k} - (k_2)_{i,j,k} \right) \frac{\tau_y}{\rho h_x} (m_1)_{i,j,k};$$

- grid equation for calculating the component of the velocity vector $w_{i,j,k}$:

$$(k_0)_{i,j,k} \frac{w^{n+\sigma} - w^n}{\tau} + \tag{6.3.54}$$

$$+(k_1)_{i,j,k} u^n_{i+1/2,j,k} \frac{(w^{n+\sigma/2}_{i+1,j,k} - w^{n+\sigma/2}_{i,j,k})}{2h_x} + (k_2)_{i,j,k} u^n_{i-1/2,j,k} \frac{(w^{n+\sigma/2}_{i,j,k} - w^{n+\sigma/2}_{i-1,j,k})}{2h_x} +$$

$$+(k_3)_{i,j,k} v^n_{i,j+1/2,k} \frac{(w^{n+\sigma/2}_{i,j+1,k} - w^{n+\sigma/2}_{i,j,k})}{2h_y} + (k_4)_{i,j,k} v^n_{i,j-1/2,k} \frac{(w^{n+\sigma/2}_{i,j,k} - w^{n+\sigma/2}_{i,j-1,k})}{2h_y} +$$

$$+(k_5)_{i,j,k} w^n_{i,j,k+1/2} \frac{(w^{n+\sigma/2}_{i,j,k+1} - w^{n+\sigma/2}_{i,j,k})}{2h_z} + (k_6)_{i,j,k} w^n_{i,j,k-1/2} \frac{(w^{n+\sigma/2}_{i,j,k} - w^{n+\sigma/2}_{i,j,k-1})}{2h_z} =$$

$$= \left((k_1)_{i,j,k} \eta_{i+1/2,j,k} \frac{w^{n+\sigma/2}_{i+1,j,k} - w^{n+\sigma/2}_{i,j,k}}{h_x^2} - (k_2)_{i,j,k} \eta_{i-1/2,j,k} \frac{w^{n+\sigma/2}_{i,j,k} - w^{n+\sigma/2}_{i-1,j,k}}{h_x^2} \right) +$$

$$+\left((k_3)_{i,j,k}\eta_{i,j+1/2,k}\frac{w^{n+\sigma/2}_{i,j+1,k}-w^{n+\sigma/2}_{i,j,k}}{h_y^2}-(k_4)_{i,j,k}\eta_{i,j-1/2,k}\frac{w^{n+\sigma/2}_{i,j,k}-w^{n+\sigma/2}_{i,j-1,k}}{h_y^2}\right)+$$

$$+\left((k_5)_{i,j,k}\eta_{i,j,k+1/2}\frac{w^{n+\sigma/2}_{i,j,k+1}-w^{n+\sigma/2}_{i,j,k}}{h_z^2}-(k_6)_{i,j,k}\eta_{i,j,k-1/2}\frac{w^{n+\sigma/2}_{i,j,k}-w^{n+\sigma/2}_{i,j,k-1}}{h_z^2}\right)+$$

$$+\left((k_5)_{i,j,k}-(k_6)_{i,j,k}\right)\frac{\tau_z}{\rho h_z}(m_1)_{i,j,k}+(k_0)_{i,j,k}g;$$

- grid equations for calculating the pressure field:

$$(k_1)_{i,j,k}\frac{P_{i+1,j,k}-P_{i,j,k}}{h_x^2}-(k_2)_{i,j,k}\frac{P_{i,j,k}-P_{i-1,j,k}}{h_x^2}+ \tag{6.3.55}$$

$$+(k_3)_{i,j,k}\frac{P_{i,j+1,k}-P_{i,j,k}}{h_y^2}-(k_4)_{i,j,k}\frac{P_{i,j,k}-P_{i,j-1,k}}{h_y^2}+$$

$$+(k_5)_{i,j,k}\frac{P_{i,j,k+1}-P_{i,j,k}}{h_z^2}-(k_6)_{i,j,k}\frac{P_{i,j,k}-P_{i,j,k-1}}{h_z^2}-\frac{\alpha(m_2)_{i,j,k}}{h_y}$$

$$\left((k_3)_{i,j,k}-(k_4)_{i,j,k}\right)==\frac{\rho}{\tau}\left((k_1)_{i,j,k}\frac{u^{n+\sigma}_{i+1/2,j,k}}{h_x}-(k_2)_{i,j,k}\frac{u^{n+\sigma}_{i-1/2,j,k}}{h_x}\right)+$$

$$+\frac{\rho}{\tau}\left((k_3)_{i,j,k}\frac{v^{n+\sigma}_{i,j+1/2,k}}{h_y}-(k_4)_{i,j,k}\frac{v^{n+\sigma}_{i,j-1/2,k}}{h_y}\right)+$$

$$+\frac{\rho}{\tau}\left((k_5)_{i,j,k}\frac{w^{n+\sigma}_{i,j,k+1/2}}{h_z}-(k_6)_{i,j,k}\frac{w^{n+\sigma}_{i,j,k-1/2}}{h_z}\right)-$$

$$-\frac{\left((k_1)_{i,j,k}-(k_2)_{i,j,k}\right)(P^{n+1}_{i,j,k}-P^{n}_{i,j,k})_{i,j,k}}{gh_x\tau^2}H\left((k_1)_{i,j,k}-(k_2)_{i,j,k}\right);$$

- grid equations for the refinement of the velocity field with respect to pressure:

$$(k_0)_{i,j,k}\frac{u_{i,j,k}^{n+1}-u_{i,j,k}^{n+\sigma}}{\tau}=-\left((k_1)_{i,j,k}\frac{P_{i+1,j,k}^{n+1}-P_{i,j,k}^{n+1}}{2h_x\rho}+(k_2)_{i,j,k}\frac{P_{i,j,k}^{n+1}-P_{i-1,j,k}^{n+1}}{2h_x\rho}\right); \tag{6.3.56}$$

$$(k_0)_{i,j,k}\frac{v_{i,j,k}^{n+1}-v_{i,j,k}^{n+\sigma}}{\tau}=-\left((k_3)_{i,j,k}\frac{P_{i,j+1,k}^{n+1}-P_{i,j,k}^{n+1}}{2h_y\rho}+(k_4)_{i,j,k}\frac{P_{i,j,k}^{n+1}-P_{i,j-1,k}^{n+1}}{2h_y\rho}\right); \tag{6.3.57}$$

$$(k_0)_{i,j,k}\frac{w_{i,j,k}^{n+1}-w_{i,j,k}^{n+\sigma}}{\tau}=-\left((k_5)_{i,j,k}\frac{P_{i,j,k+1}^{n+1}-P_{i,j,k}^{n+1}}{2h_z\rho}+(k_6)_{i,j,k}\frac{P_{i,j,k}^{n+1}-P_{i,j,k-1}^{n+1}}{2h_z\rho}\right). \tag{6.3.58}$$

Therefore, it can be noted that a finite-dimensional model of the investigated three-dimensional problem run-up of surface gravity waves is constructed, which is represented by the equations (6.3.52) ÷ (6.3.54), (6.3.55) and (6.3.56) ÷ (6.3.58).

6.4 Discrete Model Research

To maintain the conservativeness of the difference scheme, it is necessary that its grid analog satisfy the integral conservation law. The method of splitting the initial equations into physical processes is an additive scheme, it is stable and guarantees the fulfillment of the mass balance. It is established that the integral law of conservation of momentum is observed for the grid analog of the difference equation.

The approximation error of the finite-difference scheme was estimated to be $O(\tau+h_x^2+h_y^2+h_z^2)$. A stability analysis was performed for the problem using the maximum principle and limitations on time and spatial steps were obtained. The following limitations were obtained for the steps by the spatial coordinates *x*, *y*, and *z*:

$$h_x<\left\|\frac{2\mu}{u}\right\|_{C(\varpi_k)},\ h_y<\left\|\frac{2\mu}{v}\right\|_{C(\varpi_k)},\ h_z<\left\|\frac{2\eta}{w}\right\|_{C(\varpi_k)} \tag{6.4.1}$$

The stability of the discrete model was proved relative to the initial data and boundary conditions.

6.5 Results of Three-Dimensional Numerical Modeling and Their Analysis

A software code "3DbayWaves" was developed for computing three-dimensional velocity and pressure fields [Abbasov, Semenov, Tsarevsky, 2012]. The discrete equations for computing velocity components and pressure field were computed by applying an implicit scheme, while velocity components were refined using an explicit scheme.

The Algorithm of the Software Code "3DbayWaves"

The software code "3DbayWaves" was compiled and implemented in C ++ in the operating system Windows Vista. The developed program takes into account the turbulent exchange, the three-dimensional geometry of the coastal slope, friction against the bottom.

The structure of the software code 3DbayWaves consists of the following blocks:

- the input of initial data: three-dimensional geometry of the basin, the degree of water filling, wave parameters for computing the velocity and pressure, masks of the boundary conditions, and initial values of cell-filling degrees;
- the construction of grid equations for the velocity field without allowance for pressure according to the finite-volume scheme;
- the construction of grid equations for the pressure field;
- the computation of cell-filling degrees;
- the computation of the three-dimensional velocity (at the next time level);
- the computation of grid equations by applying the over relaxation method;
- output of the pressure values and the cell-filling degrees.

1. Input specification:
   - the size of the grid $\varpi_h = \{N_t\tau = T, N_x h_x = l_x, N_y h_y = l_y N_z h_z = l_z\}$;
   - N_t – upper time limit, step by time τ;

- N_x, N_y, N_z – borders on space, steps in space h_x, h_y, h_z;
- setting the three-dimensional geometry of the coastal basin;
- setting the initial level of the liquid at rest;
- initial pressure distribution on the left lateral boundary.

2. Calculation of the velocity components of the particles of the medium without allowance for pressure on the basis of the grid equations (6.3.52), (6.3.53) and (6.3.54):
   - after numbering the elements and calculating the numbers of neighboring elements in all coordinate directions, the coefficients of the grid equations for the nodes standing in the neighborhood of the center of the template;
   - calculation of the coefficients of the grid equations and right sides;
   - olution of the grid equations by an iterative method of over relaxation.
3. Calculation of the pressure field,
   - For all nodes, the coefficients and the right side of the grid equation are calculated for the calculation of the pressure according to the grid equation (6.3.55);
   - calculation of the pressure field on the basis of the SLAE calculation by the method of over relaxation.
4. Calculation of the fullness of cells.
5. Refinement of the velocity field by pressure.
6. Output of pressure field data.
7. Then increase the counter by time by one.
8. If the time is shorter than the exit time, return to step 2.
9. The end.

Structure of a Shallow Basin

Now we pass to three-dimensional simulation of the run-up of surface gravity waves. According to the formulation of the problem, a viscous incompressible fluid layer occupies a particular level of a three-dimensional coastal basin, and the bottom line passes to the coastal line. Note that the construction of marine hydraulic engineering structures of all types and classes complies with the norms and rules of Building Code. The shore-stabilizing and protecting structures include jetties, underwater breakwaters, revetments (wave heights up to 4 m), vertical structures, and groins (wave heights up to 7 m).

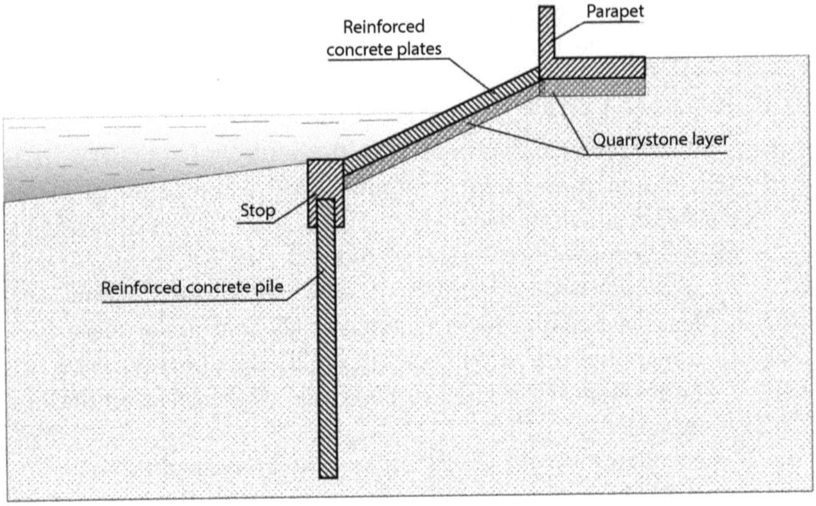

Figure 6.7.2 Schematic view of a revetment.

In view of the geomorphology of the coastal slopes (flatness and extent), the most suitable shore-protecting structures for the Sea of Azov shallow water basin are revetments [Smirnova et al., 2002]. More specifically, the structure schematically shown in Figure 6.7.2 is recommended for this goal. A revetment consists of the following structural units:

- a pebbly or sandy beach (partially flooded at run-up);
- a permeable revetment armored with reinforced concrete blocks or plates; the lower base of the revetment rests on a reinforced concrete pile cap;
- the beach berm is armored with reinforced concrete plates including a parapet (viewing platform).

Revetments with permeable blocks effectively dissipate wave energy, decrease the wave load at wave run-up, and reduce the backpressure at wave rundown.

According to construction norms, shore protection and protective structures include (taking into account the impact of waves):

- constructions with buns;
- underwater breakwaters;
- slope type constructions, wave height up to 4m;

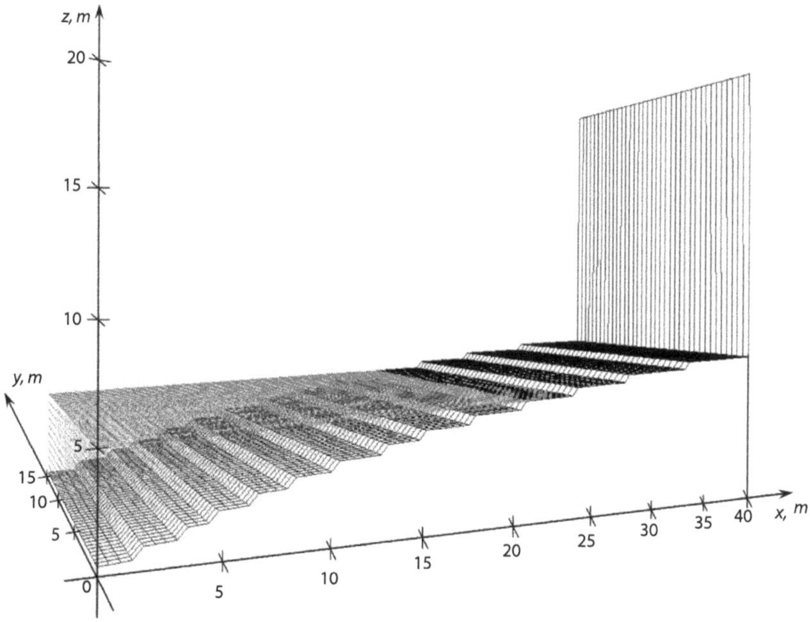

Figure 6.7.3 Three-dimensional grid of a still shallow water basin.

- vertical structures, walls, wave height up to 7m;
- mixed and special type, wave height up to 4m, length - up to 80m, water depth 10-25m.

A three-dimensional model of a coastal segment is constructed according to the chosen type of a shore-protecting structure, i.e., a revetment with a vertical wall (Figure 6.7.3.). The bottom of the basin slowly rises and passes into the coastal slope. The bathymetric conditions of the three-dimensional basin are specified according to those in Taganrog Bay of the Sea of Azov. Seawalls are usually constructed to protect seaside railroads and highways from the adverse effects of sea waves.

Results of Three-Dimensional Modeling of the Run-up

Let us turn to the results of three-dimensional numerical modeling of the run-up of nonlinear surface gravity waves on the coastal slopes of shallow water areas. According to the statement of the problem, a layer of viscous incompressible fluid fills a certain level of the three-dimensional coastal basin, while the bottom line passes to the shoreline. In accordance with the geometry of the problem, the axis Ox of the coordinate system is directed toward the shore, the axis Oy is directed along the shore, the axis Oz is directed vertically upwards.

At the initial instant of time, the liquid is at still. At a certain distance from the coast at the point $x=0$, a perturbation in the form of a pressure pulse, varying according to the harmonic law, is defined on the left lateral boundary. Let us consider the stepwise propagation of a surface wave.

The start of the 3DbayWaves code is preceded by specifying the grid sizes (N_x, N_y, N_z –are boundaries in space), the three-dimensional geometry of the coastal slope, and the number of iterations in time (N_t– upper time limit). Then the parameters of a surface gravity wave are introduced, namely, the wave amplitude – a, the wavelength –λ, and the initial level of still water – H. Next, a three-dimensional grid over the water basin is visualized as shown in Figure 6.7.3.

For the correctness of the physical process, it is necessary to observe the gradual accumulation of nonlinear effects in the propagation of the surface wave. The grid consists of 100x40x60 cells, the water surface level lies within 5÷25. The length of the bay along the Ox axis must be at least two wavelengths.

Based on the bay depth H ≤ 5m, the water segment under study is 40m long, 15m wide, and 25m deep (to the bottom), and the horizontal and vertical mesh sizes are specified as $h_x = h_y = h_z = 0.42$m.

The length of the surface wave will be $\lambda \leq 20$m, the initial values of the shallow water parameter will be in the interval $0.2 \leq kH \leq 2$, the propagation velocity of surface waves will depend on the frequency on the basis of the dispersion ratio.

Figure 6.7.4 and 6.7.5 show a sequential process of run-up a nonlinear surface gravity wave onto a gentle coastal slope. On the left lateral boundary, a harmonic perturbation is set in the form of a pressure pulse. The surface gravity wave satisfies the initial requirements of shallow water $H/\lambda < 1/5$, the length is five times the depth.

As the depth of the three-dimensional basin decreases, the wave comes onto the dry coast and the forward wave front steepens because of the enhanced nonlinear effects. Then the forward wavefront becomes sheer and breaks. This process can be observed in Figure 6.7.4b. The breaking wave is scattered and collides with the forward vertical wall of the basin (Figure 6.7.4c).

Next, the water mass flows back down the coastal slope and wave run-down occurs (Figure 6.7.5a). The backpropagating wave hits the crest of the next wave, steepening the latter and promoting its breaking. By the arrival of the next wave crest, the steel water level has risen (Figure 6.7.5c).

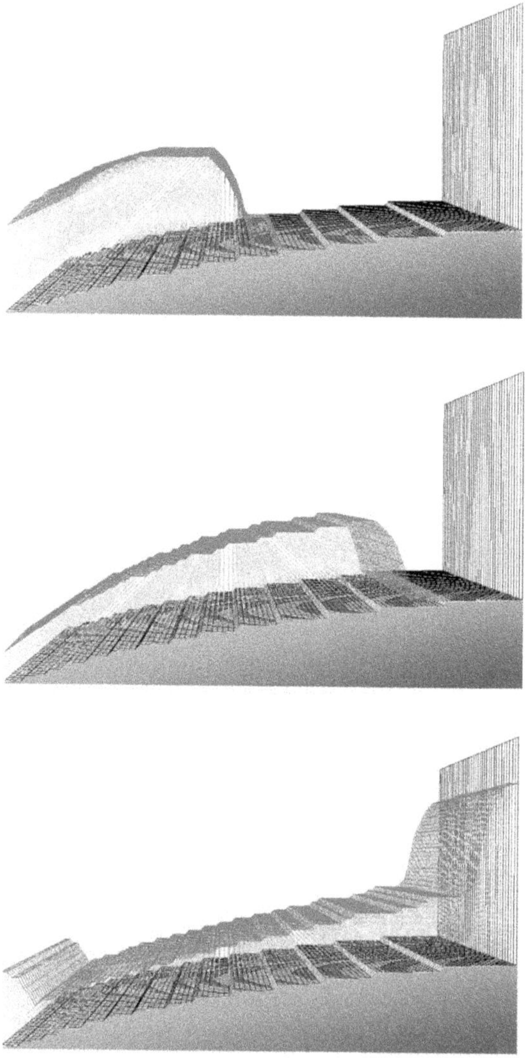

(a)

(b)

(c)

Figure 6.7.4 Consecutive stages of the run-up of a surface gravity wave with initial parameters: $f = 0.13$Hz; $\lambda = 50$m; $c = 6.6$m/s; H = 5m; $a = 4.2$m; $kH = 0.6$; $\varepsilon = 0.8$; at various times (a) $t = 3.7$s; (b) $t = 4.3$s; (c) $t = 6.3$s.

These run-up stages are also observed for gravity surface waves with other initial shallow water parameters.

Figure 6.7.6 shows the process of run-up surface gravity waves with different initial parameters of shallow water, for which the propagation patterns described above are also preserved.

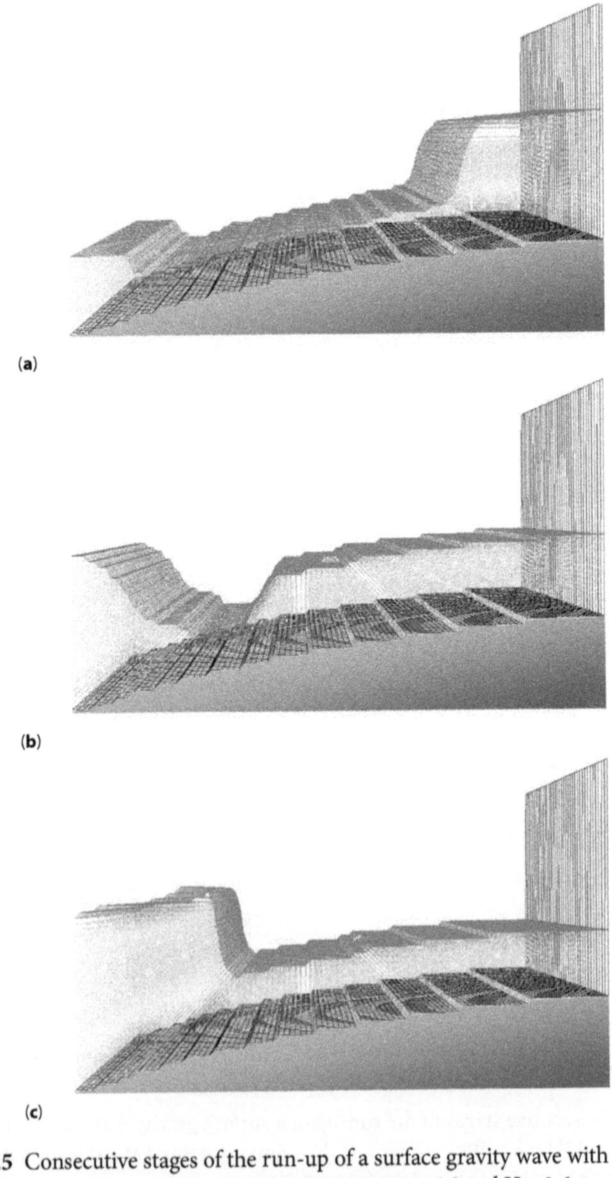

Figure 6.7.5 Consecutive stages of the run-up of a surface gravity wave with initial parameters: $f = 0.13$Hz; $\lambda = 50$m; $c = 6.6$m/s; H = 5m; $a = 4.2$m; $kH = 0.6$; $\varepsilon = 0.8$; at various times (a) $t = 7.7$s; (b) $t = 9.9$s; (c) $t = 10.8$s.

To verify the stability of the model developed, we consider the run-up of surface waves in another shallow basin, namely, 40m long, 10m wide, and 20m deep. The grid consists: 40x10x20 cells; the water surface level lies within 5÷15 cells in the vertical; the horizontal and vertical mesh sizes are specified as $h_x = h_y = h_z = 1.0$m.

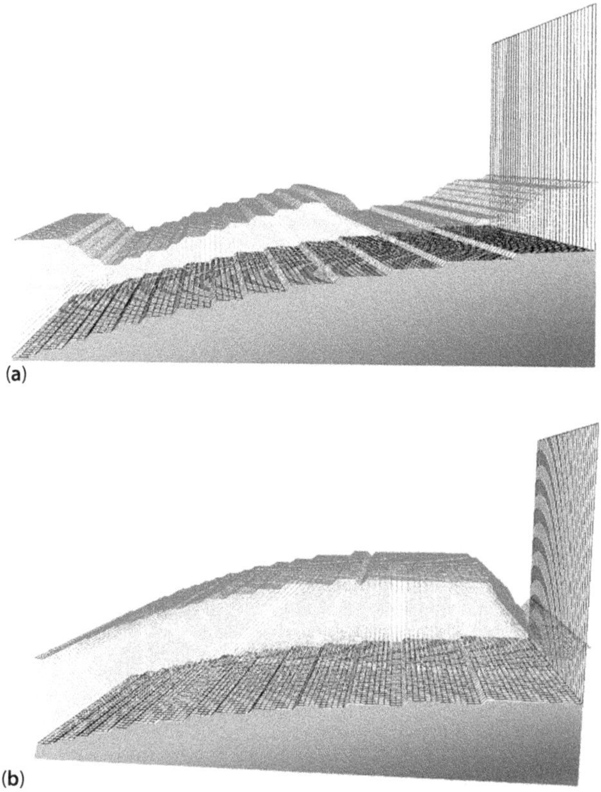
(a)

(b)

Figure 6.7.6 Consecutive stages of the run-up of a surface gravity wave with initial of shallow water parameters: (a) $f = 0.25$Hz; $\lambda = 25$m; $c = 6.2$m/s; H = 8.4m; $a = 2.5$m; $kH = 2.1$; $\varepsilon = 0.3$; $t = 8.8$s; (b) $f = 0.12$Hz; $\lambda = 75$m; $c = 8.9$m/s; H = 10m; $a = 2.5$m; $kH = 0.8$; $\varepsilon = 0.2$; $t = 5.7$s.

Figure 6.7.7 and Figure 6.7.8 illustrated the consecutive stages in the run-up of a nonlinear surface gravity wave with the following initial parameters: wavelength $\lambda = 60$ m; amplitude $a = 5.5$ m; depth H = 5 m (the basin grid is not shown). An analysis of the run-up process, it can be noted that the wave processes are again preserved. The wave comes onto the dry coast, the forward front steepens and breaks, and the wave recedes from the vertical wall.

It should be stressed that the run-up patterns represented in Figures 6.7.5 ÷ 6.7.8 demonstrate the spatial distribution of the water mass pressure inside the shallow basin. Using these distributions, we can estimate the force action of waves on coastal structures. The pressure value is determined by the color of the filled cell. The color scale ranges from red to dark blue, and the cells with maximum pressure values are shown in red.

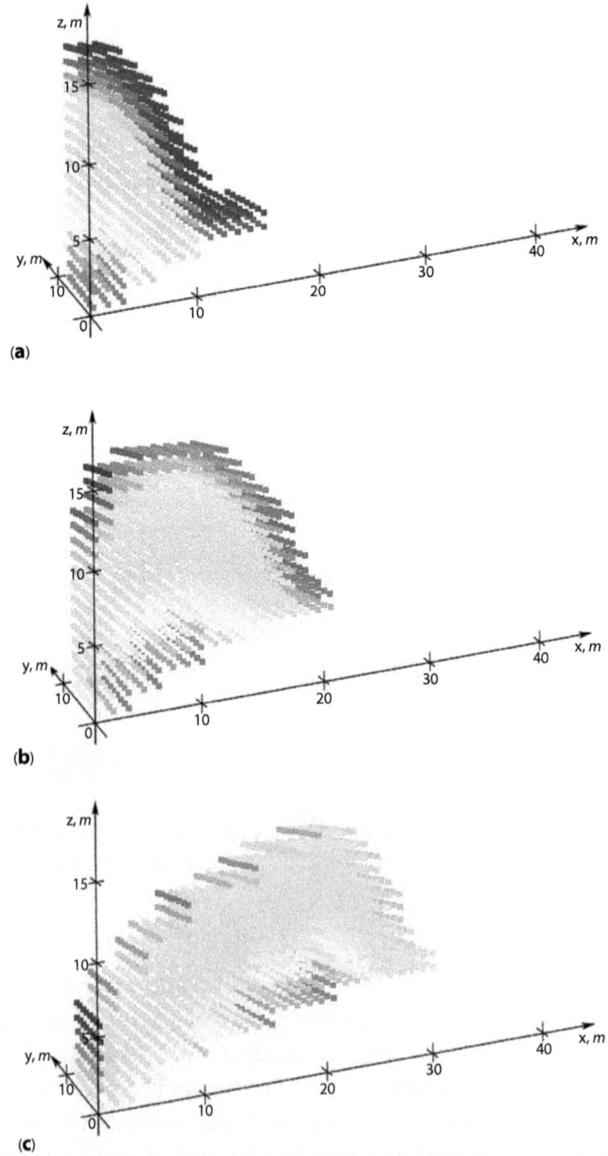

Figure 6.7.7 Consecutive stages of the run-up of a surface gravity wave with initial parameters: f = 0.12Hz; λ = 60m; c = 6.7m/s; H = 5m; a = 5.5m; kH = 0.5; ε = 1.1; at various times t: (a) 3.1s; (b) 4.8s; (c) 6.9s.

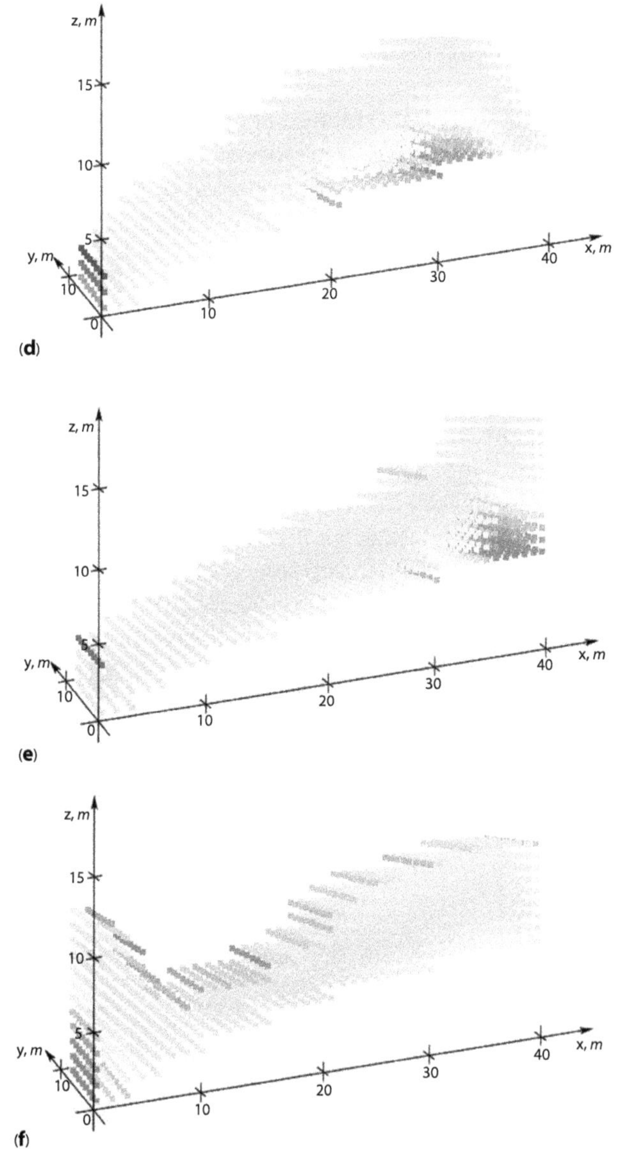

Figure 6.7.8 Consecutive stages of the run-up of a surface gravity wave with initial parameters: $f = 0.12$Hz; $\lambda = 60$m; $c = 6.7$m/s; H = 5m; $a = 5.5$m; $kH = 0.5$; $\varepsilon = 1.1$; at various times t: (d) 9.7s; (e) 12.7c; (f) 20.9s.

For larger water column heights (higher hydrostatic pressure) and higher particle velocities (hydrodynamic pressure), the influence exerted on the basin walls is stronger.

In Figure 6.7.9, the spatial distribution of the water mass pressure inside the basin at the run-up of surface gravitational waves is displayed in color. This distribution provides a detailed description of the force action produced by the surface wave not only on the basin boundaries, but also over the entire internal volume.

In the case under consideration, for a wave with – f=0.12Hz; λ=60m; c=6.7m/s; a=5.5m the hydrodynamic pressure p=53.8kPa and the velocity of the particles u=4.15m/s on the water surface at the right wall of the basin are: the run-up height is 12m. These parameters correspond to an incoming surface wave at the breaking stage. For the same wave with a smaller amplitude (steepness): a=2.25m, we have u=1.65m/s, and p=22kPa, and the run-up height is 3m. As the column height increases, the influence exerted by the hydrostatic pressure is enhanced as well. The strongest action (Figure 6.7.9) at the maximum run-up stage occurs at its base. Next, run-down begins and the load on the wall is weakened (until the next run-up).

The developed model makes it possible to assess the force impacts not only on the surface of the shore protection structure, but throughout the shallow water area. Analyzing these results, we note that the physical processes proceeding in the two-dimensional case are overall preserved in the three-dimensional run-up simulation. However, since the complexity

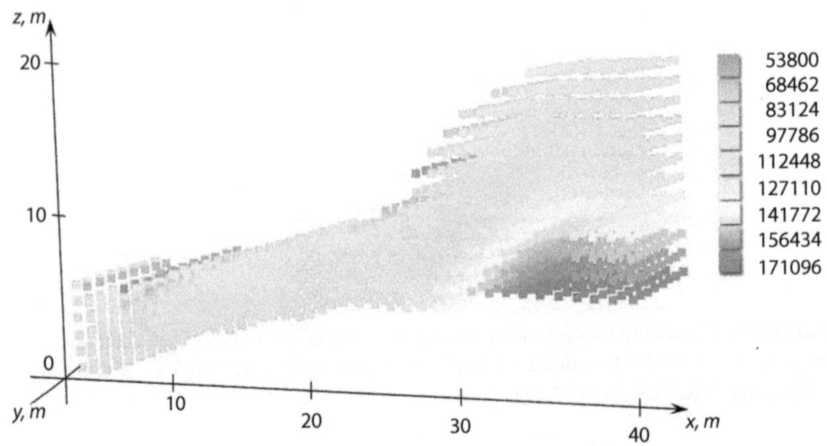

Figure 6.7.9 Indication (color scale) of the water mass pressure inside the basin for the run-up of a surface gravity wave with f = 0.12Hz; λ = 60m; c = 6.7m/s; H = 5m; a = 5.5m; kH = 0.5; ε = 1.1; t = 16.4s.

of the computations increases, the specification of the basin geometry becomes more complicated. For this reason, a surface gravity wave of high initial steepness was used to obtain a more pronounced run-up process.

6.6 Discussion and Comparison of Results

Let's compare the results of three-dimensional numerical simulation of the run-up of nonlinear surface gravity waves on the coastal slopes with the results of numerical studies by other authors. Figure 6.8.1 presents the results of two models for three-dimensional numerical modeling of the run-up of nonlinear surface gravity waves.

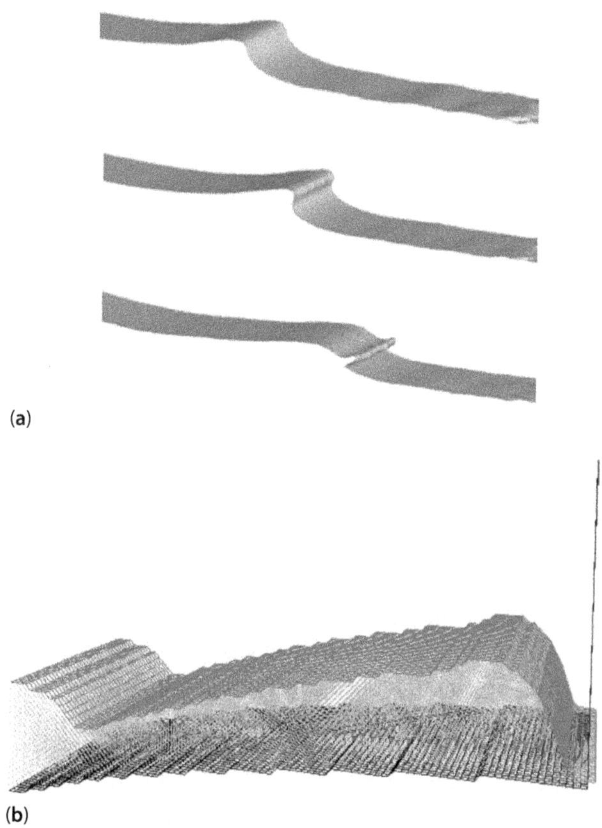

Figure 6.8.1 Comparison of the process of run-up of a surface gravity wave on the basis of different models: (a) $\lambda = 2.5$m; H = 0.7m; $T = 1.3$s; $a = 0.07$m [Lubin et al., 2006]; (b) $\lambda = 25$m; H = 6.7m; $c = 7.5$m/s; $a = 4.2$m; $t = 4.7c$ (developed model).

Figure 6.8.1a shows the results of simulation of surface wave run-up onto a sloping shore on the basis of the Navier-Stokes equations within the two-phase medium [Lubin et al., 2006]. Figure 6.8.1b shows the process of run-up the surface gravity wave along the coastal slope on the basis of our model.

As a result of the comparison, we can note:

- for both models the surface wave profile begins to be distorted due to the influence of depth, the forefront crest slope begins to be steepened;
- later, the wave breaks with the formation of a curl on the crests, since the two-phase air-water interaction is considered (Figure 6.8.1a), within our model, the breaking of the crests occurs without the formation of a curl (Figure 6.8.1b).

The comparison revealed that the results agree well in the basic stages of wave propagation and breaking. It should be noted that the three-dimensional model can also be used to estimate the force actions produced by wave processes on shore-protecting structures in shallow water basins.

Conclusions

In the sixth chapter the questions of 3D computational modeling of run-up of nonlinear surface gravity waves based on Navier-Stokes equations were examined.

The task statement was defined, the task geometry was provided, the boundary and initial conditions were described. Using the physical process split method, a discrete model on a uniform grid was created. Mesh schemas of the original differential equations was received using the integral-interpolation method.

A discrete, finite-volume model of the task to be investigated, based on the factor of a cell filling developed. The discrete analogues of the equations were obtained to calculate the components of the velocity vector, the pressure field.

A study of the conservativeness of a discrete model was carried out, and an approximation of a finite differential was found that had the first order of time and the second order in three spatial coordinates.

The grid equations were solved by the over relaxation method. A 3DbayWaves program code has been developed to calculate the 3D velocity and pressure field of the aquatic environment, and a description of the structure and algorithm of the program were presented. The structure of the 3D basin takes into account the particularities of the coastal formations of the Azov Sea, and an assessment of the force's effects on the coast protection works was made.

The results of 3D computational modeling of the run-up of nonlinear surface gravity waves to the coast formations of shallow water are provided. The 3D visualization of the step-by-step run-up of the surface wave to the coast slope was presented. The behavior of the surface gravity waves with various initial parameters of shallowness is considered, for which the revealed features of the propagation process are also preserved. To verify the correctness of the model, the results of the 3D computational modeling were compared to the existing computational data.

The results of the studies in this chapter have been published in the following main works: [Abbasov, 2015 Lenand], [Abbasov, Semenov, Tsarevsky, 2012-2], [Abbasov, 2014-1], [Abbasov, 2014-2].

Conclusion

The first chapter presented the features of problems of mathematical physics, the classification of the linear differential equations of the second order to describe these objectives, the nonlinear equations of hydrodynamics and the methods of solving them, and considered the application of the method of successive approximations.

The second chapter carried out research and modeling of wave activities on the shallow water surface. The main equations that describe the surface gravity waves, main properties of the waves at shoaling are provided. The graphs of the main characteristics of the surface gravity wave from the depth and from the wave number in the shallow water were provided.

The 3D models of the propagation of the group of surface waves in shallow water are presented in the linear case. The 3D models of coastal formations were created in the form of cape, bay and straight coastline. The 3D models of the refraction of the group of surface gravity waves at are presented at various coast formations in the linear case.

The waves on the surface of the fluid that emerged when the solid body was moving on its surface were considered. The basic parameters of a wave tail behind a moving object were described. In-kind illustrations of wave activities on the shallow water surface were presented: wave ripples and swells, refractions of waves in bay conditions, distribution of ship waves.

In the third chapter, a study and modeling of nonlinear wave activities on the shallow water surface was carried out. In order to describe the surface gravity waves at shallow water, the equations of shallow water by Boussinesq, Korteweg-de Vries were examined. Using the method of successive approximations, two approximations of the shallow water equation were obtained, taking into account the quadratic nonlinearity. Using the proposed graphic method the constraints for the horizontal velocity of particles in the primary and secondary waves were found.

The profiles of nonlinear surface waves were presented for the various travel distances and the energy performance diagrams. 3D models of coast formations were made, the 3D models of distribution and refraction the surface gravity waves.

The reasons for the asymmetry of surface wave profiles have been considered. The appearance of high-frequency harmonics of the main wave leads to sharper crest and a flat fronts. In-kind illustrations of the sharpness of the crest of surface waves, the formation of diver breaker, and the breaking of the coastline were presented.

Based on analytical expressions the impacts of dispersion effects to waves profile transformation are taken into account. The 3D distribution and refraction models of the nonlinear surface wave in the bay, based on dispersion, were presented.

In the fourth chapter the questions of modeling the distribution of surface gravity waves based on computational solution of the equation of shallow water were examined. The results of the numerical velocity calculation and the free surface elevation function for the different values of the initial steepness and the depth of the shallow water were shown. The refraction of computational modeling of the run-up of nonlinear surface gravity waves at beach approach of various slopes are modeled.

In the fifth chapter for 2D computational modeling of the run-up process of the nonlinear surface gravity waves the equations of Navier-Stokes were used. The results of the 2D computational modeling of the run-up process of the nonlinear surface gravity waves based on the Navier-Stokes equations were received and analyzed. To adequately describe the physical process, the effects of the bottom geometry on the wave run-up process have been examined.

In the sixth chapter the questions of 3D computational modeling of run-up of nonlinear surface gravity waves based on Navier-Stokes equations are described. The construction of the coast protection work of the slope type for the conditions of the shallow water model was proposed and an assessment of the force's effects on it was made. The results of 3D computational modeling the run-up of the surface gravity waves were provided and analyzed.

The results of analytical and computational research of the modeling the distribution and run-up of nonlinear surface gravity waves at shallow water could be used for the monitoring and forecasting of wave activities, their impact to sediment and suspension transportation process.

The results of 3D computational modeling of the distribution of the nonlinear surface gravity waves at shallow water could be used for ranging the force impact to coast protection works, for the brow detection at the submarine relief, and shore riff detection based on wave pattern at the water surface.

References

1. Abbasov I.B. *Scattering of nonlinearly interacting acoustic waves: sphere, cylinder, spheroid*. Moscow: Fizmatlit, 2007, 160p.
2. Abbasov I.B. *Modeling of nonlinear wave phenomena on the shallow water surface*. Moscow: Fizmatlit, 2010, 128p.
3. Abbasov I.B. *Three-dimensional numerical simulation the run-up of nonlinear surface gravity waves*. Moscow: Lenand, 2015, 160p.
4. Abbasov I.B. Spatial modeling of wave phenomena on the surface of a gulf. // Izvestiya Vuzov. North-Caucasian region. Natural series. 2001. №4. P.56–57.
5. Abbasov I.B. Study and simulation of nonlinear surface gravity waves under shallow-water conditions //Izvestiya, *Atmospheric and Oceanic Physics*. V.39, №4, 2003, P.506–511.
6. Abbasov I.B. Investigation and spatial modeling of nonlinear surface gravity waves in shallow water //Izvestiya Vuzov. North-Caucasian region. *Natural Sciences*. 2003. №1.P.33–36.
7. Abbasov I.B. Investigation and modeling of the refraction of nonlinear surface waves in a gulf//Izvestiya, *Atmospheric and Oceanic Physics*.V.40, №3, 2004, P.377–380.
8. Abbasov I.B. Simulation of the spatial refraction of the non-linear surface gravitational waves. VI International Congress on Mathematical Modeling. Book of abstracts. Nizhny Novgorod, University of Nizhny Novgorod. 2004. P.143.
9. Abbasov I.B. Modelling of nonlinear surface gravity waves in shallow water with regard to dispersion//*Doklady Earth Sciences*. 2009. V.429. №9. P.1605–1607, doi:10.1134/S1028334X09090414
10. Abbasov I.B. Transformation of nonlinear surface gravity waves under shallow-water conditions //*Applied Mathematics*. 2010. V.1. №4. P.260–264. doi:10.4236/am.2010.14032
11. Abbasov I.B. Modelling of nonlinear surface gravity waves under shallow-water conditions with account of dispersion //*Waves in Random and Complex Media*. 2011. V.21. №1. P.13–22. doi.org/10.1080/17455030.2010.500702
12. Abbasov I.B. Numerical simulation of nonlinear surface gravity waves transformation under shallow-water conditions //*Applied Mathematics*. 2012. V.3. №2. P. 135–141 doi:10.4236/am.2012.32021

13. Abbasov I.B., Neverov A.A. Program Code of modeling of surface wave propagation in shallow water "BayWaves". The certificate №2202616207 on the state registration of the computer program, was filed on May 22, 2012, registered on 06.07.2012.
14. Abbasov I.B., Semenov I.S., Tsarevsky V.V. Program Code of two-dimensional simulation of surface waves in shallow water "2DBayWaves". The certificate №2012616206 on the state registration of the computer program, was filed on May 22, 2012, registered on 06.07.2012.
15. Abbasov I.B., Semenov I.S., Tsarevsky V.V. Program Code of three-dimensional simulation of surface waves in shallow water "3DBayWaves". Certificate №2012617087 on the state registration of the computer program, was filed on May 22, 2012, registered on 08.08.2012.
16. Abbasov I.B. Mathematical modeling the runup of nonlinear surface gravity waves. //*International Journal of Pure and Applied Mathematics*. 2013. V.82. №1. P.133–142.
17. Abbasov I.B. Simulation the runup of nonlinear surface gravity waves a steep coastal slope // *International Journal of Pure and Applied Mathematics*. 2013. V.84. №3. P.299–306. doi:10.12732/ijpam.v84i3.15
18. Abbasov I.B. Three-dimensional simulation the runup of nonlinear surface gravity waves on shallow coast //*International Journal of Pure and Applied Mathematics*. 2014. V.93. №1. P.135–145. doi:10.12732/ijpam.v93i1.11
19. Abbasov I.B. Three dimensional simulation of the runup of nonlinear surface gravity waves //*Computational Mathematics and Mathematical Physics*. 2014. V.54, №5, P. 900–914. doi:10.1134/S0965542514050030
20. Adalsteinsson, D., Sethian, J.A. Transport and Diffusion of Material Quantities on Propagating Interfaces via Level Set Methods //*J. Comp. Phys*, 2002. V.185, №1, P.271–288
21. Afanasiev K.E., Stukolov S.V. Numerical simulation of the work of the experimental generator of single surface waves //*Bulletin of the Kemerovo State University*. 2013. № 3–3 (55). P.6–14.
22. Airy G.B. Tides and Waves. *Encyclopedia Metropolitana*. 1845. London. v.5. P.241–396.
23. Aleshkov Yu.Z. *Mathematical modeling of physical processes*. Publishing of St. Petersburg State University. St. Petersburg, 2001, 264p.
24. Aleshkov Yu.Z. Transformation of surface waves in a liquid with a slowly varying depth. Sat. //*Actual problems of continuum mechanics*, Leningrad State University, №13, 1980. P.34–38.
25. Amosov A.A., Dubinsky Yu.A., Kopchenova N.V. *Computational methods for engineers*. Moscow: Higher School, 1994. 544p.
26. Amsden A. A., Harlow F. H. A simplied MAC technique for incompressible fluid flowcalculations //*J. Comp. Phys*. 1970. V.6. P.322–325
27. Aramanovich I.G., Levin V.I. *Equations of mathematical physics*. M.: Science. Ch. Ed. Fizmatlit. 1969. 288p.

28. Arkhipov D.G., Khabakhpashev G.A. A new approach to the description of spatial nonlinear waves in dispersive media //*Reports of the Academy of Sciences*. 2006. T.409. №4. P.476–480.
29. Bakhtyar R., Barry D.A., Yeganeh-Bakhtiary A., Ghaheri A. Numerical simulation of surf-swash zone motions and turbulent flow //*Adv. Water Resour.* 2009. №32. P.250–63.
30. Bakhvalov N.S., Zhidkov N.P., Kobelkov G.M. *Numerical methods*. 4th ed. Moscow: BINOM. Laboratory of Knowledge, 2006. - 636p.
31. Barakhnin V.B., Khakimzyanov G.S. On the algorithm for the numerical solution of equations of a nonlinear dispersive shallow water model // *Computational technologies*. 1996. V.1.№3. P.5–19.
32. Beizel S.A., Khakimzyanov G.S., Chubarov L.B. Modeling of surface waves generated by an underwater landslide moving along a spatially inhomogeneous slope //*Computational technologies*. 2010. V.15. №3. P.39–51
33. Belotserkovsky O.M. *Numerical modeling in the mechanics of continuous media*: 2nd ed., Moscow: Fizmatlit, 1994. 448p.
34. Belotserkovsky O.M., Davydov Yu. M., Nonstationary method of "large particles" for gasdynamic computations //*Journal of Computational Mathematics and Mathematical Physics*. 1971. V.11, P.182–207.
35. Belotserkovsky O.M., Gushchin V.A., Kon'shin V.N. Splitting method for investigating stratified fluid flows with a free surface //*Journal of Computational Mathematics and Mathematical Physics*. 1987. V.27. №4. P.594–609.
36. Benjamin T.B., Feir J.E. The disintegration of wave trains on deep water //*J. Fluid Mech.* 27. 1967. P.417.
37. Boussinesq J. *Theorie de l'intumescence liquide appelee onde solitaire ou de translation, se propageant dans un canal rectangulaire*. C.R.Acad. Sci., Paris. 1871. 72. P.755–759.
38. Borisova N.M. On the modeling of the process of the onset of a discontinuous wave on an inclined shore //*Siberian Journal of Calculated Mathematics*, 2007. V.10, №1. P.1–18.
39. Borisova N.M., Ostapenko V.V. On numerical modeling of the process of propagation of discontinuous waves along a dry bed //*Journal of Computational Mathematics and Mathematical Physics*. 2006. Vol. 46, No. 7, P.1322–1344.
40. Brekhovskikh L.M., Goncharov V.V. *Introduction to the mechanics of continuous media*. Moscow: Nauka, 1982. 335p.
41. Bretherton F. P. Resonant interactions between waves. The case of discrete oscillations //*J.Fluid Mech.*, 1964, v. 20, p. 457–479.
42. Bystrov Yu.A., Isaev S.A., Kudryavtsev N.A., Leontiev A.I. *Numerical modeling of heat transfer in stacks of pipes*. SPb.: Shipbuilding, 2005. 392p.
43. Caponi E. A., Saffman P. G., Yuen H. C. Instability and confined chaos in a nonlinear dispersion wave system. *Phys. Fluids*, 1982.v. 25, №. 12, p.2159–2166.

44. Cauchy A.. Theorie de la propagation des ondes a la surface d'un fluide pesant d'une profondeur indefine. *Oeuvres Completes d'Augustin Cauchy*. I serie. 1815. P.5–318.
45. Chapalain G., Cointe R., Temperville A. Observed and modeled resonantly interacting progressive water-waves. *Coastal Eng*. 1992. 16. P. 267–300.
46. Chen Y., Guza R.T., Elgar S. Modeling spectra of breaking surface waves in shallow water //J. Geophys. Res. 1997. V.102. №15. P.35–46
47. Chikin A.L. Three-dimensional problem of calculation of hydrodynamics of the Sea of Azov //*Mathematical Modeling*, 2001. V.13, №2, P.86–92.
48. Chorin A. J. A numerical method for solving incompressible viscous flow problems //*J. Comput. Phys*. 1967. V.2, №1. P.12–26.
49. Chorin A. Numerical solution of the Navier-Stokes equations //*Math. Comp*. - 1968. Vol. 22. – P. 745–762.
50. Cokelet E.D. Steep gravity waves in water of arbitrary uniform depth. *Philos. Trans. R. Soc. London, Ser. A*., 1977, v. 286, P.183–230.
51. Colagrossi A., Landrini M. Numerical simulation of interfacial flows by Smoothed Particle Hydrodynamics //*J. Comput.Phys*. 2003. V.191. P.448–475.
52. Debol'skaya E. I., Yakushev E.V., Sukhinov A.I. Formation of Fish Kills and Anaerobic Conditions in the Sea of Azov//*Water resources*. 2005. V.32, №.2. P.151–162.
53. Delis A.I., Kazolea M. Finite volume simulations for the generation and propagation of long waves //*Proceeding 5th European Congress on Computational Methods in Applied Sciences and Engineeering (ECCOMAS 2008) June 30 – July 5, 2008, Venice, Italy*.
54. Delis A., Skeels C.P. TVD schemes for open channel flow //*Int. J. Numerical Methods in Fluids*. 1998. V.26. P. 791–809.
55. Didenkulov O.I., Didenkulova I.I., Pelinovsky E.N. Parametrization of the characteristics of the rolling of single waves in a bay of parabolic cross section //*Proceedings of NSTU*. R.E. Alekseev. 2014. №3. P.30–36.
56. Didenkulova I.I., Pelinovsky E.N. Long-wave rolling on the shore: the influence of the shape of a suitable wave //*Oceanology*. V.48. №1. 2008. P.5–10.
57. Didenkulova I.I., Zaibo N., Kurkin A.A., Pelinovsky E.N. Steepness and spectrum of a nonlinearly deformed wave in shallow water //Izvestiya RAN. *Physics of the Atmosphere and the Ocean* Vol. 42, №6, 2006, P.839–842.
58. Douglas J., Gunn J. E. A general formulation of alternating direction implicit methods. Pt. 1. Parabolic and hyperbolic problems //*Numer. Math*. 1964. V. 6, №5. P.428–453.
59. Dubreil-Jacotin M.L. Sur les theoremes d'existence relatives aux ondes permanents periodiques a deux dimensions dans les liquids heterogenes //*J. Math. Pures Appl*. 1937. 16. 9. P.43–67.
60. Dwight G.B. *Tables of integrals and other mathematical formulas*. M.: Science. 1983. 176p.
61. *Dynamics of channel flows and lithodynamics of the coastal zone of the sea*. Ed. VC. Debolsky. M.: Science. 1994. - 303p.

62. Eldeberky Y., Madsen P.A. Determenistic and stochastic evolution equations for fully dispersive and weakly nonlinear waves //*Coastal Eng.*, 38. 1999. P. 1–24.
63. Elgar S., Norheim C. A., Herbers T. H. Nonlinear evolution of surface wave spectra on a beach //*Journal of Physical Oceanography*. 1998. V.28. №7. P.1534–1551.
64. Euler L. Principes generaux du movement des fluides. Mem. Acad. Berlin.1955. 11. P.274–315. ("Opera Omnia", 1755. *Ser. Secunda*.Vol. XII. P. 54–91).
65. Fedotova Z.I. Justification of the numerical method for simulating the rolling of long waves ashore //*Computational technologies*. 2002. T.7. № 5. P.58–76.
66. Fedotova Z.I. On a system of nonlinear-dispersion equations of hydrodynamics with a useful property //*Computational technologies*. 2004. V.9. №6. P.82–88.
67. Fedotova Z.I., Khakimzyanov G.S. Nonlinear dispersion equations of shallow water on a rotating sphere //*Computational technologies*. 2010. V.15. № 3. P.135–145.
68. Fletcher K., *Computational methods in the dynamics of liquids*. Part 2. - M.: Mir, 1991. 552p.
69. Flick R.E., Guza R.T., Inman D.L. Elevation and velocity measurements of laboratory shoaling waves //*J. Geophys. Res*. 1981. V. 86. № C5. P. 4149–4160.
70. Gabov S.A. *Introduction to the theory of nonlinear waves*. Moscow: Moscow State University. 1988. 176p.
71. Gabov S.A. On the property of destruction of solitary waves, described by the Whitham equation //*Dokl. RAN*, 1979. V.246. №6. P.1292–1295.
72. Gardner C.S., Greene J.M., Kruskal M.D., Miura R.M. Method for solving the Korteweg-de Vries equation//*Phys. Rev. Lett*. 1967. V.19. P.1095–1097.
73. Gerstner. Theorie der Wellen. Abh. d.k. bohm. Ges. d. Wiss. 1802. (Gilbert's Annalen der Physik, XXXII. 1809).
74. Gill A. *Dynamics of the Atmosphere and the Ocean*. In two volumes, V.1. M.: Mir. 1986. 396p. V.2. 415p.
75. Gimenez G., Chamant M., Farnand J.P. Non-linear response of a single bubble driven by a two-components exciting wave //*Proc. 10-th Int. Symp. on nonlinear Acoustics, Kobe, Japan. 1984*. P. 83–87.
76. Goda Y., Morinobu K. Breaking wave heights on horizontal bed affected by approach slope //*Coastal Engineering Journal*. 1998. V. 40. № 4. P.307–326.
78. Gusev O.I., Shokina N.Yu., Kutergin V.A., Khakimzyanov G.S. Modeling of surface waves generated by an underwater landslide in the reservoir // *Computational technologies*. 2013. T.18, № 5, P.74–90.
79. Guyer R.A., Johnson P.A. Nonlinear mesoscopic elasticity evidence for a new class of materials //*Physics Today*. 1999. V.52. №4. P.30–36.
80. Haa T., Shima J., Linb P., Cho Y.-S. Three-dimensional numerical simulation of solitary wave run-up using the IB method //*Coastal Engineering*. V.84, №2. 2014, P.38–55.

81. Harlow F. H. Welch J. E. Numerical calculation of time-dependent viscouse incompressible flow of fluid with free surface //*Phys. Fluids*. 1965.V.8, №12. P.2182–2189.
82. Harlow F. Numerical method of particles in cells for hydrodynamic problems. *Computational methods in hydrodynamics*. - M.: Mir. - 1967. P. 316 -342.
83. Haslett R. determination of the acoustic scatter patterns and cross sections offish models and ellipsoids //*Brit. Journ. Appl. Phys*. 1962.V.13. №12. P.611–620.
84. Hasselmann K. On the nonlinear energy transfer in gravity-wave spectrum. General theory. *J. Fluid Mech*., 1962,V.12, P.481–500.
85. Hasselmann K. Weak-interaction theory of ocean waves. In: *Basic developments in fluid dynamics*. V.2. New York: Academic Press, 1968, P.117–182.
86. Havelock T.H. The propagation of groups of waves indispersive media, with application to waves on water produced by a travelling disturbance. *Proc. Roy. Soc. Lond*. (A). 1908. P.398–430.
87. Helmholtz H. *Theorie der Luftschwingungen in Rohren mit offenen Enden*, Crelle, LVII, 14.1859.
88. Herbers T. H., Elgar S., Sarap N.A., Guza R.T. Nonlinear dispersion of surface gravity waves in shallow water //*Journal of physical oceanography*. 2002. V.32. №4. P. 182–1193.
89. Hirt C. W., Nichols B. D. Volume of fluid (VOF) method for the dynamics of free boundaries //*J. Comp. Phys*. 1981. V.39. P201–225.
90. Hodgins D.O., Le Blond P.H. and Huntley D.A. Shallow-water wave calculations //*Canadian Contractor Report of Hydrography and Ocean Sciences*. 1985. №10. 75p.
91. Hogner E. A contribution to the theory of ship-waves. Ark. *Math. Astr. Fys*. 17.V.12. 1922. P.1–50.
92. Gonorovsky I.S. *Fundamentals of radio engineering*. M., 1957. 727p.
93. Hydrometeorology and hydrochemistry of the USSR seas. V.V. Sea of Azov. S.-Pb.: Gidrometeoizdat. 1991. P.75–88.
94. Infeld E., Rowlands J. *Nonlinear waves, solitons and chaos*. M.: Fizmatlit, 2006. 480p.
95. Jeffreys H. On water waves near the shore //*Phil. Mag*. 1924. 48. P.44.
96. Kadomtsev B.B., Petviashvili V.I. On the stability of solitons in media with a weak dispersion //*Reports of the Academy of Sciences*. 1970. V.192. P.753–756.
97. Kalmykov V.A. Numerical simulation of collapsing waves //*Matem. modeling*. 1996. V. 8. №8. P.37–41.
98. Kawasaki K. Numerical simulation of breaking and post-breaking wave deformation process around a submerged breakwater //*Coastal Engineering Journal*/ 1999. V.41. №3&4. P.201–223
99. Kawasaki K., Takasu Y., Ut H.D. 2-d Numerical wave flume with solid-gas-liquid interaction and its application //*Coastal Engineering*, 2010. 57, P.1–15.

100. Kelvin (W. Thomson). Deep sea Ship-waves. 1910. *Mathematical and Physical Papers*. V.4. Cambridge, P.394–418.
101. Khakimzyanov G.S., Shokina N.Yu. Numerical simulation of steady-state fluid flows within the shallow water model //*Computational technologies*. 1996. V.1. № 3. P.93–105.
102. Kimmoun O., Branger H. 2007. A PIV investigation on laboratory surf-zone breaking waves over a sloping beach //*J. Fluid Mech*, 588, P.353–397.
103. Kogan V.R., Kuznetsov V.V. Application of the theory of analytic functions in the numerical simulation of nonstationary surface waves //*Journal of Computational Mathematics and Mathematical Physics*. 1995. V. 36. № 10. P.1448–1456.
104. Korn G., Korn T. *Handbook of Mathematics*. M., 1968. 720p.
105. Korteweg D.J., de Vries G. On the change form of long waves advancing in a rectangular channel and on new type of long stationary waves.//*Phyl.Mag*. 1895. 5. P. 422–443.
106. Krasitsky V.P. To the theory of transformation of the spectrum in the refraction of wind waves //Izv. AN SSSR. *Physics of the atmosphere and ocean*. 1974. V.10. №1. P.72–82.
107. Krukier L.A. Mathematical modeling of hydrodynamics of the Sea of Azov during the implementation of projects for the reconstruction of its ecosystem //*Mathematical Modeling*, 1991. V.3, №9, P.3–20.
108. Krukier L.A., Chikin A.L., Chikina L.G., Shabas I.N. *Modeling of hydrophysical processes in water bodies with extensive areas of shallow-dyas*. SFU publishing house. Rostov-on-Don, 2009, 244p.
109. Krukier L.A., Muratova G.V., Andreeva E.M., Zubov V.N., Subbotina T.N., Nikitenko O.B. *Mathematical modeling of the processes of convective-diffusion transport in problems of ecology*. SFU publishing house. Rostov-on-Don, 2008, 304p.
110. Krylov Yu. M. To the theory of refraction of sea waves. *Proceedings of the GOIN*. 1950. Issue. 16. P.95.
111. Kshevetskii S. P., Vereshchagina I.S. Numerical modelling of tsunami wave run-up and breaking within a two-dimensional atmosphere–ocean two-layer model //*Natural Hazards and Earth System Sciences*, №2, 2014, P.3397–3421.
112. Kudryashov N.A., Sytsko Yu.I., Chesnokov S.A. Mathematical modeling of gravitational waves in the ocean in the approximation of "shallow water" // *Pis'ma Zh*. 2003. №77. V.10. P.649.
113. Kudryashov N.A. Non-linear waves and solitons //*Soros' Educational Journal*. 1997. №2. P.85–91.
114. Kulikovsky A.G., Pogorelov N.V., Semenov A.Yu. *Mathematical problems of numerical solution of hyperbolic systems of equations*. Moscow: Fizmatlit, 2001. 608p.
115. Kulikovsky A.G., Reutov V.A. Propagation of nonlinear waves over semi-infinite underwater depressions and ridges //Iz. AN SSSR. *Mechanics of fluid and gas*. 1980. № 2. P.53–61.

116. Kushnir V.M., Dushko V.R., Kramar V.A. Influence of surface gravity waves on coastal oceanic structures //*East-European Journal of Advanced Technologies*. 2013. V.6. №5. P.36–41.
117. Kuznetsov S.Yu., Saprykina Y.V. Experimental studies of the evolution of wave groups in the coastal zone of the sea //*Oceanology*. 2002. V. 42, №3, P.356–363.
118. Kuznetsov S.Yu., Saprykina Ya.V., Cherneva Zh., Andreeva N. Spatial-temporal variability of the amplitude-phase structure of storm waves in the coastal zone of the sea //*Oceanology*. 2009. №4, P.206–215.
119. Lagrange J.L. *Memoire sur le theorie du mouvements des fluides*. 1781. Nouv. Mem. Acad. R. Sci. Bellelett. (Reprinted in *Oeuvres*, Vol. IV. P. 695–750. Gauthier-Villars. Paris. 1869).
120. Lamb G. *Hydrodynamics*. Moscow, Gostekhizdat, 1947. 524p.
121. Landsberg G.S. *Elementary textbook of physics*. Moscow: Nauka, 1964. V. 3. 572p.
122. Laplace P.S. *Recherches sur plusieurs point du systeme du monde*. 1778. Mem. Acad. R. Sci. Paris. P.75–182. (*Oeuvres* Vol. 9. P. 71–183. Gauthier-Villars, Paris. 1893).
123. Lavrentiev M.A., Shabat B.V. *Problems of hydrodynamics and their mathematical models*. M.: Science. 1973. 416p.
124. Lax P. Integrals of non-linear equations of evolution and solitary waves // Com. *Pure and Appl. Math*., 1968, v. 21, P.467–490.
125. Lee Ch., Liu J., Lee Ch., Kang Y. The Second-Order KdV Equation and its Soliton-Like Solution //Modern Physics Letters B. 2009. V.23, Issue 14, P.1771–1780
126. Leontyev I.O. *Coastal dynamics: waves, currents, sediment flows*. Moscow: GEOS, 2001. 272p.
127. Levi-Civita T. Determinazione rigrosa della onde irrotationale periodiche in acqua profonda //*Atti. Accad. Lincei*. 1924. 33. 5. P.141–144.
128. Lighthill M. J. Some special cases treated by the Whitham theory. *Proc. R. Soc. London*, Ser. A, 1967, v. 299, P.28.
129. Lighthill M.J. *River waves*. Naval hydrodynamics publication. 515 National Academy of Sciences. National Research Council. 1957.
130. Litvinenko A.A., Khabakhpashev G.A. Numerical modeling of nonlinear sufficiently long two-dimensional waves on water in basins with a shallow bottom //*Computational technologies*. 1999. V. 4, №3, P.95–105.
131. Liu P.H., Losada I.J. Wave propagation modeling in coastal engineering // *Journal of Hydraulic Research*. 2002. V.40. №3. P.229–240.
132. Loitsyan L.G. *Mechanics of Fluids and Gas*. M.: Nauka, 1978. 736p.
133. Longuet-Higgins M. S. The instabilities of gravity waves of finite amplitude in deep water. II. Subharmonics. *Proc. R. Soc. London Ser*. A, 1978, v. 360, P.489–505.
134. Longuet-Higgins M.S., Smith N.D. An experiment on third order resonant wave interactions //*J. Fluid Mech*. 25. 1966. P.417–436.

135. Lonquet-Higgins M.S. Grest instabilities of gravity waves //*J. Fluid Mech.*, 1993.
136. Lowan A., Morse P., Feshbach H., Lax M. Scattering and radiation from circular cylinder and spheres, mathematical tables. US Navy Department, Washington. 1946.
137. Lubin, P., S. Vincent, S. Abadie and J.-P. Caltagirone. Three-dimensional Large Eddy Simulation of air entrainment under plunging breaking waves // *Coastal Engineering*, 2006. V.53, P.631–655.
138. Lubin, P., Vincent, S., Caltagirone, J.-P., Abadie, S. Three-dimensional large eddy simulation of vortices induced by plunging breaking waves //*Proceeding ASCE 29th International Conference on Coastal Engineering*, 2004. P.253–262
139. Mamykina V.A., Khrustalev Yu.P. *Coastal zone of the Azov Sea*. Rostov n / D. RSU Publishing House. 1980. 176p.
140. Marchuk A.G. The method for calculating the rolling of long gravitational waves on a sloping bank //Evolution of a tsunami from the source to the shore. M.: *Radio and Communication*, 1982. P.64–68.
141. Marchuk A.G., Moshkalev P.S. Numerical simulation of tsunami waves on the shore of an arbitrary profile //Vestn. Novosib. state. University. Series: Information technology. 2014. V.12, № 2. P.55–63.
142. Marchuk G.I. *Methods of splitting*. Moscow: Nauka, 1988. 264p.
143. Michell J.H. The highest waves in water //*Phil. Mag*. 1893.v. 36. 5. P.430–437.
144. Monin A.S., Krasitsky V.P. *Phenomena on the surface of the ocean*. L.:Gidrometeoizdat. 1985. 375p.
145. Morse F., Feshbach G. *Methods of theoretical physics*. M.: Izd. In. lit. 1960, 896p.
146. Muratova G.A., Krukier L.A., Nikitenko O.B., Chikin A.L., Shabas I.N. Modeling of hydrophysical processes in the reservoir. In *Complex monitoring of the environment and biota of the Azov basin*, 2004 Apatity, v.VI, P.279–198.
147. Naumkin P.N., Shishmarev I.A. On the existence and destruction of waves described by the Whitham equation //Dokl. RAN. 1986. - V.288. №1. P.90–95.
148. Nekrasov A.I. *On waves of steady-state species* //Izvestiya Ivanovo-Voznesensk Polytechnic Institute. 1921. 3. P. 52–65.
149. Newton J.S. *Philosophiae Naturalis Principia Mathematica*. Londini, Anno MDCLXXXVII. Julii 1686.
150. *Nonlinear waves*. Ed. S. Leibovich and A. Seabass. Moscow: Mir. 1977. 319p.
151. Ostrovsky L.A., Pelinovsky E.N. Transformation of waves on the surface of a fluid of variable depth. Izv. AN SSSR. *Physics of the atmosphere and ocean*. 1970. V.6. № 9. P.934–939.
152. Patancar S. V., Spolding P. V. Calculation Procedure for Heat, Mass, and Momentum Transfer in Three-dimensional Parabolic Flows //*Int. J. Heat and Mass Transfer*. 1972.V.15. P.1787–1806.
153. Peaceman D. W., Rachford H. H. The numerical solution of parabolic and elliptic differentional equations //*J. Soc. Indust. Appl. Math*. 1955. V. 3, №1. P.28–41.

154. Peregrine D.H. Interaction of water waves and currents //*Adv. Appl. Mech.* 1976. V. 16. P. 9–117.
155. Peregrine D.H. Long waves on a beach //*J. Fluid Mech.* 1967. V.27. part IV. P.815–827
156. Peregrine D.H., Banner M.L. Wave breaking in deep water //*Ann. Rev. Fluid Mech.*, 1993. V.25. P. 373–397.
157. Petters A.S. A new treatment of the ship-wave problem //*Comm. Pure. Appl. Math.* 2. V.2–3. 1949. P. 123–148.
158. Phillips O.M. *The Dynamics of the Upper Ocean.* Cambridge: Univ. Press, 1966.
159. Poisson S. D. *Memoire sur la Theorie des Ondes.* Mem. de l'Acad. Roy. des Sciences.1816.
160. Prudnikov A.P., Brychkov Yu.A., Marichev O.I. *Integrals and series.* M.:Science. 1981. 800p.
161. Rankine W.J. On the Exact Form of Waves near the Surface of Deep Water // *Phil. Trans.* 1863. P.481.
162. Rayleigh (J.W. Strutt). On waves. *Phill. Mag.* 1876. 5. 1. P.257–279.
163. Richthmayer R., Morton K. *Difference methods for solving boundary value problems.* M.: Mir, 1972. 421p.
164. Richtmyer R., Morton K. *Difference methods for initial-value problems.* New York: Wiley, 2nd edn. 1967. 309p.
165. Riemann B. *Uber die Fortpflanzung ebener Luftwellen von endlicher Schwingungsweite, Gottingen Abbandlungen.* Bd. VIII, S. 43, Werke, 2te Aufl., Leipzig, 1892, P.157.
166. Roach P. *Computational Fluid Dynamics.* M.: Mir, 1980. 616p.
167. Roseau M. Contribution a la theorie des ondes liquids de gravite en profondeur variable. *Publ. Sci. Tech. Du Ministere de l'Air.* 1952. Paris. 275. P.90.
168. Rumyantsev B.N. On unsteady motions of a heavy fluid near an inclined shore //*PMM.* 1960. V.24. №3. P.554–557.
169. Russell J. Scott. Report on waves. *Br. Assoc. Adv. Sci. Rep*, 1844, 14th Mtg, p. 311–390, plates XLVII-LVII.
170. Samarsky A.A. *The theory of difference schemes.* - Moscow: Nauka, 1989. 348p.
171. Sekier-Zenkovich Ya.I. To the three-dimensional problem of standing waves of finite amplitude on the surface of a heavy liquid //*DAN SSSR.* 1952. V.86. №1. P.35–38.
172. Selezov I.T. *Modeling of wave and diffraction processes in continuous media.* Kiev. Science. dumka. 1989. 204p.
173. Selezov I.T., Sidorchuk V.N., Yakovlev V.V. *Transformation of waves in the coastal zone of the shelf.* Kiev. Science. dumka. 1982. 208p.
174. Selezov I.T., Tkalich P.V. Nonlinear refractive model of the propagation of surface waves in a liquid //*Lavrentiev Readings on Mathematics, Mechanics and Physics. II All-Union Conf. Kiev.* 1985. P.187–189.

175. Shermeneva M.A., Shugan I.V. Calculation of the rolling of waves on a gentle slope //*Letters in ZhTF*. 2006. V.32. № 2. P.33–38.
176. Shimozono T., Sato S. A Study on nonlinear dispersive wave equations and wave breaking solution //*Journal of Japan Society of Civil Engineers*, Ser. B2 (Coastal Engineering). 2009. V. B2–65. №1. P.1–5.
177. Shlyamin B.A. Excitement in the Sea of Azov //Tr. state. oceanographer. Institute. 1977. V.16. P.45–47.
178. Shokin Y.I., Chubarov L.B., Marchuk A.N., Simonov K.V. *Computational experiment in the tsunami problem*. Novosibirsk: Science. Sib. Otd, 1989. 168p.
179. Shuleikin V.V. *Physics of the sea*. M., Publishing "Science". 1968. 1061p.
180. Shuleikin V.V. Refraction of waves on the continental shelf //*Izv. AN SSSR*. Omen. 1935. №10. P.1355.
181. Shuleikin V.V. The destruction of waves under the action of shallow water // *Izv. AN SSSR*. 1954. № 1. P.65.
182. Shup Terry, E. *Applied numerical methods in physics and engineering*. M.: Higher School, 1990. 254p.
183. Smirnova T.G., Pravdivets Yu. P., Smirnov G.N. *Coastal structures*. Textbook. for universities. - Moscow: Publishing ASV, 2002. - 302p.
184. Sretensky L.N. Periodic waves generated by a source above an inclined bottom //*DAN SSSR*. 1963. V.151. №5. P.1050–1052.
185. Sretensky L.N. *Theory of wave motions of a fluid*. The main edition of the physical and mathematical literature of the publishing house "Nauka". M., 1977. 816p.
186. Stokes G.G. On the theory of oscillatory waves //*Trans. Camb. Phil. Soc*. 8. 1847. 441p.
187. Struik D.J. Sur les ondes irrotationelles dans les canaux //*Atti. Accad. Lincei*. 1925. 16. 1. P.522–527.
188. Sukhinov A.I. Additive schemes for modeling three-dimensional heat conduction equations in cylindrical and spherical coordinates //*Differential equations*. 1987, V. 23. №12. P.2122–2132.
189. Sukhinov A.I. Precision Mathematical Models of Small Water Bodies // *Mathematical Modeling*. 2003. №10. P.17–34.
190. Sukhinov A.I., Timofeeva E.F., Chistyakov A.E. Construction and investigation of a discrete mathematical model for the calculation of coastal wave processes //Izvestiya SFU. *Technical science*. - 2011. №8. P.22–32.
191. Sukhinov A.I. Calculation of the coefficient of vertical turbulent exchange for models of shallow water bodies // *Mathematical Modeling and Information Technologies*. South-Russian. state. tech. un-t (NPI). Novocherkassk. Ed. journal. "Izv. universities. Electromechanics ", 2007. P.72–76.
192. Sukhov A.I. Complex oceanological studies of the Azov Sea in the 28th voyage of the research vessel Akvanavt //*Oceanology*, 2003. V.43. №1. P.44–53.
193. Tikhonov A.N., Samarsky A.A. *Equations of mathematical physics*. Moscow: Nauka, 1966. 724p.

194. Ting F.C.K., Kirby J.T. Dynamics of surf-zone turbulence in a spilling breaker //*Coastal Engineering*, 1996. V.27. P.131–160.
195. Touboula J., Pelinovsky E. Bottom pressure distribution under a solitonic wave reflecting on a vertical wall //*European Journal of Mechanics - B/Fluids*. 2014, V.48, №6. P.13–18.
196. Ursell F. On Kelwin's ship-wave pattern //*J. Fluid Mech*. 8. V.3. 1960. P. 418–431.
197. Ursell F. On the rigorous foundations of short-wave asymptotics. *Proc. Cambr. Phill. Soc*. 1966. 62. 2. P.227–244.
198. Vilenkin N.Ya. *The method of successive approximations*. M.: Science. Home ed. physical and mathematical. letter. 1968. 108p.
199. Vinje T., Brevik P. Numerical simulation of breaking waves //*Adv. Water Resour*. 1981, №4. P.77–82.
200. Vinogradova M.B., Rudenko O.V., Sukhorukov A.P. *Theory of waves*. M.:Science. Home ed. physical and mathematical. letter. 1979. 384p.
201. Viottia C., Carbonea F., Dias F. Conditions for extreme wave runup on a vertical barrier by nonlinear dispersion //*Journal of Fluid Mechanics*. 2014, V.748, №6. P.768–788.
202. Vlasova E.A., Zarubin V.S., Kuvyrkin G.N. *Approximate methods of mathematical physics*. Moscow: Izd. MSTU Bauman. 2001. 700p.
203. Voltsinger N.E., Pyaskovsky R.V. *The theory of shallow water. Oceanological problems and numerical methods*. L.: Gidrometeoizdat. 1977. 207p.
204. Watanabe Y., Saeki H. Three dimensional large eddy simulation of breaking waves //*Coastal Engineering Journal*, 1999. V.41, № 3–4, P.281–301.
205. Whitham G.B. Nonlinear dispersion of water waves //*J. Fluid Mech*., 1967, v.27, P.399–412.
206. Whitham G.B. A general approach to linear and nonlinear dispersive waves using a Lagrangian. *J. Fluid Mech*., 1965, v. 22, P. 273–283.
207. Whitham G. *Linear and nonlinear waves*. Mir, Moscow, 1976. 622p.
208. Wu Y.-T., Y. C.-L., H. Shih-Chun. Three-dimensional numerical simulation on the interaction of solitary waves and porous breakwaters //*Coastal Engineering*.V.85, №3. 2014. P.12–29.
209. Yuen G., Lake B., *Non-linear dynamics of gravitational waves on water*. M.: Mir, 1987. 179p.
210. Yuen H. C, Lake B. M. Nonlinear deep water waves: Theory and experiment. *Phys Fluids*, 1975, v 18, P. 956–960.
211. Zabusky N.J., Kruskal M.D. Interaction of solitons in a collisionless plasma and the recurrence of initial states //*Phys. Rev. Lett*. 1965, V.15, P.240–243.
212. Zaharov V. Weakly non-linear waves on the surface of an ideal finite depth fluid //*Amer. Math. Soc. Trans*. 1998. V. 182. P.167–197.
213. Zakharov V.E. Stability of periodic waves of finite amplitude on the surface of a deep liquid //*Applied Mechanics and Theoretical Physics*. 1968. № 2. P. 86–94.

214. Zakharov V.E., Shabat A.B. The exact theory of two-dimensional self-focusing and one-dimensional self-modulation of waves in nonlinear media //*Zh. Eksp. Fiz.*, 1971. V.74, P.118–134.
215. Zeytunyan R. Nonlinear long waves on the water surface and solitons // *Uspekhi Fizicheskikh Nauk*. 1995. V.165. № 12. P.1403–1456.
216. Zhao Q., Armfield S., Tanimoto K. Numerical simulation of breaking waves by a multi-scale turbulence model //*Coastal Engineering*, 2004. V.51. P.53–80
217. Zheleznyak M.K., Pelinovsky E.N. Physicomathematical models of the tsunami downwind on the shore //*The tsunami on the beach. Collection of scientific works*. Gorky, 1985. P.8–34.
218. Zhilyaev A.P. Calculation of fluctuations in the level of the Azov Sea // *Oceanology*. 1972. V.12. №1. P.49–56.

Index

Lightning Source UK Ltd.
Milton Keynes UK
UKHW021950120219
337194UK00003B/76/P

9 781119 487968